Robust Image Authentication in the Presence of Noise

Nataša Živić
Editor

Robust Image Authentication in the Presence of Noise

Springer

Editor
Nataša Živić
University of Siegen
Siegen
Germany

ISBN 978-3-319-13155-9   ISBN 978-3-319-13156-6 (eBook)
DOI 10.1007/978-3-319-13156-6

Library of Congress Control Number: 2015935937

Springer Cham Heidelberg New York Dordrecht London
© Springer International Publishing Switzerland 2015
This work is subject to copyright. All rights are reserved by the Publisher, whether the whole or part of the material is concerned, specifically the rights of translation, reprinting, reuse of illustrations, recitation, broadcasting, reproduction on microfilms or in any other physical way, and transmission or information storage and retrieval, electronic adaptation, computer software, or by similar or dissimilar methodology now known or hereafter developed.
The use of general descriptive names, registered names, trademarks, service marks, etc. in this publication does not imply, even in the absence of a specific statement, that such names are exempt from the relevant protective laws and regulations and therefore free for general use.
The publisher, the authors and the editors are safe to assume that the advice and information in this book are believed to be true and accurate at the date of publication. Neither the publisher nor the authors or the editors give a warranty, express or implied, with respect to the material contained herein or for any errors or omissions that may have been made.

Printed on acid-free paper

Springer is part of Springer Science+Business Media (www.springer.com)

Preface

"A picture is worth a thousand words," is a well-known proverb by Confucius. Images help in the communication of ideas, events or memories very efficiently and effectively.

Data communications involve transmission over noisy channels. Due to the nature of modern communications, intentional modifications or unintentional errors might be introduced into images by noise during the transmission over a communication channel, a human or by source encoders and decoders. In order to prevent intentional modifications, images are protected more and more against manipulations by authentication mechanisms. Most of the unintentional errors can be corrected using channel coding, but the complete elimination of all the errors cannot be ensured. Such erroneous images cannot be processed, as they are recognized as untrustworthy by authentication mechanisms.

This fact raises lots of questions and expresses a need for new algorithms which support authentication of images in the *presence of noise*, i.e., unintentional modifications of images. Standard authentication mechanisms are extremely fragile to the existence of errors and therefore they very often fail in communications in noisy environments. Therefore, *robustness* is an important property which new algorithms have to support.

This book consists of six chapters presenting different aspects of its main contribution, which is the introduction of the robustness into image authentication in noisy communications.

The advancement of digital technology implicated the development of sophisticated tools to tamper digital images. Therefore, a number of techniques using digital image watermarking and hashing have been proposed. In Chapter 1, the problem of image authentication is introduced and a wavelet-based technique is presented to address robustness, security and tamper detection issues in hash-based image authentication schemes. Unlike other hashing schemes that apply randomness to select image features from a known feature space, a randomized pixel modulation method is used to randomly transform the entire feature space. The key dependent transformed feature space is used to calculate the image hash. To reduce the size of the hash, 3-bit and 4-bit quantization schemes are also presented. A number of experimental results along with analysis are reported to show how an image hashing scheme is practically evaluated to gauge its robustness, security and tamper detection capability.

Chapter 2 is concerned with the digital multimedia content which is susceptible to malicious manipulations and alterations. There are two widely used image authentication techniques, i.e., digital signatures and digital watermarking. Digital signatures based authentication is quite mature and already in wide use. However, its shortcoming is that the signature needs extra transmission capacity or establishing a separate secure channel for transmission. Additionally, due to the usage of secure hash functions, digital signatures are also susceptible to failed authentication because of the avalanche effect, which might occur due to many reasons such as channel noise, quantization or compression. Digital watermarking can perform content authentication without the aforementioned shortcomings and it is difficult to be detached or tampered. An overview of digital watermarking techniques is given in this chapter, followed by the state of the art and the introduction of a new algorithm.

Chapter 3 investigates the feasibility of content based image authentication using the perceptual image hashing technique as an alternative data authentication scheme in Wireless Multimedia Sensor Networks (WMSNs). High level requirements for image authentication in WMSNs are addressed, and the previously published literature regarding data authentication schemes for WMSNs as well as content based image authentication using the perceptual image hashing technique is discussed as well. Furthermore, the performance of five selected perceptual image hashing algorithms are measured and compared in terms of robustness, discriminability and security in order to provide a perspective on the potential feasibility as an alternative solution to the existing data authentication scheme in WMSNs.

Chapter 4 gives a brief and explicit survey on generic approximate (fuzzy) message authentication codes. The presented schemes are described and categorized according to their design methods. The main design methods are based on computational complexity and unconditional security viewpoints. The corresponding analysis of each category is given, including the performance and security consideration. A short comparison result is given at the end of the first part of the chapter. The second part shows some applications of generic approximate message authentication codes on image authentication techniques.

Chapter 5 addresses the sensitivity of standard message authentication codes (MACs) which makes them unsuitable for application in multimedia applications. Specific algorithms for image authentication, so called fuzzy image authentication algorithms, are discussed in this chapter. These algorithms are tolerant to a certain degree of modifications introduced by channel noise or image processing operations. Fuzzy image authentication algorithms should be designed in a manner that they can differentiate allowed manipulations from manipulations which are not allowed. The fuzzy authentication algorithms should have an additional desirable property of error localization and correction. Once the modifications in the image are localized to a specific region within the image, they can be corrected using the error correcting codes embedded in the authentication algorithms.

The robustness of biometric systems under affine transformations is important for precise user authentication. Most systems are using alignment to compensate for occurring transformations. Chapter 6 analyses the robustness of alignment-free biometric features in fingerprint and vein biometrics, motivated by the drawbacks of alignment-based biometrics. Various approaches from the fields of pattern recognition as well as biometrics are described, aggregated and assessed. Based on the similarities of the involved disciplines, an evaluation strategy from the field of digital image processing is proposed to assess the performance and invariance of biometrical feature extraction.

The future communications will be developed in the direction of integrated and secure techniques which need to be fast and reliable. The rapid progress specifically in wireless communications needs solutions for the biggest enemy of communications, i.e., noise. The solutions which can support such development trends have to be robust against noise. This book presents and discusses some algorithms which contribute to an improved authentication of images subjected to noise from different sources.

Contents

1	**Hash-Based Authentication of Digital Images in Noisy Channels**		1
	Fawad Ahmed and Amir Anees		
	1.1	Introduction	1
		1.1.1 Basic Requirements of Data Security	2
		1.1.2 Multimedia Security	2
		1.1.3 Image Hash	4
		1.1.4 Basic Requirements of Image Hashing	5
	1.2	Review of Image Hashing Schemes	6
		1.2.1 Statistic-Based Schemes	6
		1.2.2 Relation-Based Schemes	7
		1.2.3 Coarse Representation-Based Schemes	8
		1.2.4 Matrix-Based Schemes	9
		1.2.5 Low-Level Feature-Based Schemes	10
	1.3	Illustration of an Image Hashing Scheme	11
		1.3.1 Hash Generation Module	12
		1.3.2 Image Verification Module	14
	1.4	Reducing Size of Image Hash Using Quantization	15
		1.4.1 Hash Generation Using 3-Bit Quantization	15
		1.4.2 Image Verification Using 3-Bit Quantization	17
		1.4.3 3-Bit Quantization Example	18
		1.4.4 A 4-Bit Quantization Scheme	21
		1.4.4.1 Hash Generation Using 4-Bit Quantization	21
		1.4.4.2 Image Verification Using 4-Bit Quantization	23
		1.4.4.3 Reduction in Hash Size Due to 4-Bit Quantization	25
		1.4.4.4 Effect on Robustness and Tamper Detection Due to 4-Bit Quantization	25
	1.5	Performance Evaluation	26
		1.5.1 The Effect of Beta (β) on Hash Collision	27
		1.5.2 Robustness to Channel Noise and JPEG Compression	28
		1.5.3 Threshold Selection	30
		1.5.4 Detection of Tampering	31

	1.5.5	The Receiver Operating Characteristic Curve	32
	1.5.6	Hash Size	33
1.6	Security Analysis		33
	1.6.1	The Impact of Randomized Pixel Modulation on System's Security	34
	1.6.2	Statistical Analysis	36
	1.6.3	Effect of Secret Key on the Hash	37
	1.6.4	Probability of Hash Collision	38
	1.6.5	Quantization and System's Security	39
1.7	Conclusion		40
References			41

2 Watermarking for Image Authentication ... 43
Chen Ling and Obaid Ur-Rehman

2.1	The Basis of Watermarking		44
2.2	Classification		46
	2.2.1	Perceptibility	47
	2.2.2	Detection Types	48
	2.2.3	System Platform	49
	2.2.4	Image Compression	49
	2.2.5	Embedding Domain	49
		2.2.5.1 LSB	50
		2.2.5.2 SS	50
		2.2.5.3 DCT	53
		2.2.5.4 DWT	54
	2.2.6	Robustness	54
	2.2.7	Lossless	56
2.3	Requirements of Watermarking		57
	2.3.1	Fidelity	57
	2.3.2	Capacity	58
	2.3.3	Robustness	58
	2.3.4	Security	58
		2.3.4.1 Malicious Attacks	59
		2.3.4.2 Incidental Attack	61
		2.3.4.3 Secret Key	61
	2.3.5	Tamper Detection, Location, and Recovery	62
2.4	A Watermarking Algorithm for Image Authentication		64
	2.4.1	Watermark Design	64
		2.4.1.1 Feature Extraction	66
		2.4.1.2 Watermark Embedding Approach	69
	2.4.2	An Example Design Case	69
		2.4.2.1 Goals	69
		2.4.2.2 Hybrid Feature Watermark Generation	69

| | | 2.4.2.3 | Watermark Embedding | 71 |
| | | 2.4.2.4 | Watermark Extraction and Authentication Procedure | 71 |

References 72

3 Perceptual Image Hashing Technique for Image Authentication in WMSNs ... 75
Jinse Shin and Christoph Ruland

- 3.1 Introduction 75
- 3.2 High Level Requirements for Image Authentication in WMSNs ... 77
- 3.3 Previous Works on Data Authentication in WMSNs 80
- 3.4 Perceptual Image Hashing 81
  - 3.4.1 Basic Concept 81
  - 3.4.2 Desirable Properties 82
    - 3.4.2.1 Perceptual Robustness 82
    - 3.4.2.2 Fragility to Visual Distinct Image 83
    - 3.4.2.3 Unpredictability of the Hash 83
- 3.5 Content Based Image Authentication Using Perceptual Image Hashing Technique 83
  - 3.5.1 Image Statistics Based Approach 85
  - 3.5.2 Relation Based Approach 87
  - 3.5.3 Coarse Image Representation Based Approach 88
  - 3.5.4 Low-level Image Representation Based Approach 89
- 3.6 Experiment Results 91
  - 3.6.1 Robustness 92
  - 3.6.2 Discriminability 95
  - 3.6.3 Security 97
- 3.7 Conclusion 98

References 102

4 A Review of Approximate Message Authentication Codes 105
S. Amir Hossein Tabatabaei and Nataša Živić

- 4.1 Introduction 105
  - 4.1.1 Definitions and Notations 106
- 4.2 Dedicated AMAC Schemes 107
  - 4.2.1 Majority-Based AMACs 107
    - 4.2.1.1 Shifting Attack 108
    - 4.2.1.2 Security Enhancement 109
    - 4.2.1.3 Analysis of the Majority-Based AMACs 110
  - 4.2.2 Noise Tolerant Message Authentication Codes (NTMACs) . 112
    - 4.2.2.1 Analysis of the NTMAC-Based AMACs 115
  - 4.2.3 AMACs Based on Computational Security 117
    - 4.2.3.1 Analysis of the Computational Security Based AMACs 119
  - 4.2.4 Unconditionally Secure AMAC 120

		4.2.4.1 Analysis of Unconditionally Secure AMAC	121
	4.2.5	Comparison	121
4.3	Applications of Dedicated AMACs in Image Authentication Techniques		122
	4.3.1	Extension of the MAJORITY-Based AMACs and the NTMACs	122
	4.3.2	Extension of $AMAC_1$ and $AMAC_2$	123
4.4	Conclusion		125
References			126

5 Fuzzy Image Authentication with Error Localization and Correction — 129
Obaid Ur-Rehman and Nataša Živić

- 5.1 Introduction … 129
- 5.2 Building Blocks of the Fuzzy Image Authentication Algorithms … 131
  - 5.2.1 Content Based Authentication … 131
  - 5.2.2 Discrete Cosine Transform … 131
  - 5.2.3 Error Correcting Codes … 132
    - 5.2.3.1 Error Correcting Codes in Image Authentication … 132
    - 5.2.3.2 Reed–Solomon Codes … 132
    - 5.2.3.3 Turbo Codes … 133
- 5.3 Applications of Error Correcting Codes in Image Authentication … 134
- 5.4 Fuzzy Image Authentication Codes with Error Localization and Correction … 135
  - 5.4.1 Fuzzy Authentication Based on Image Features … 135
  - 5.4.2 Image Error Correcting Column-Wise Message Authentication Code (IECC-MAC) … 136
  - 5.4.3 Image Error Correction Noise Tolerant Message Authentication Code (IEC-NTMAC) … 141
- 5.5 Performance and Security Analysis of the Proposed Fuzzy Authentication Algorithms … 146
  - 5.5.1 Performance Study … 146
  - 5.5.2 Security Analysis … 147
- 5.6 Simulation Results … 148
  - 5.6.1 Simulation Parameters … 148
  - 5.6.2 Data Rate Analysis … 149
  - 5.6.3 Simulation Results for IECC-MAC … 150
  - 5.6.4 Simulation Results Using IEC-NTMAC … 152
  - 5.6.5 Image Error Rate (IER) … 152
- References … 153

6 Robustness of Biometrics by Image Processing Technology — 155
Robin Fay and Christoph Ruland

- 6.1 Introduction … 155
- 6.2 Pattern Recognition and Biometric Authentication … 156
  - 6.2.1 Pattern Recognition Systems … 156

	6.2.2	Affine Invariant Pattern Recognition	157
		6.2.2.1 The Problem of Affine Invariant Pattern Recognition	157
		6.2.2.2 Alignment-Free Features	158
	6.2.3	Biometric Authentication	161
		6.2.3.1 Biometric Systems	161
		6.2.3.2 Template Protection	163
		6.2.3.3 Fingerprint Recognition	164
		6.2.3.4 Vein Pattern Recognition	165
6.3	Feature Extraction in Fingerprint Biometrics		167
	6.3.1	Global Features	167
	6.3.2	Minutiae Based Feature Extraction	168
		6.3.2.1 General	168
		6.3.2.2 Minutiae Detection	168
		6.3.2.3 Minutiae Description	170
	6.3.3	Image Based Feature Extraction	171
		6.3.3.1 Image Based Feature Detection	171
		6.3.3.2 Image-Based Feature Description	173
6.4	A Comparison of Alignment-Free Biometric Systems		173
	6.4.1	Global Methods	173
	6.4.2	Image Based Methods	174
	6.4.3	Minutiae-Based Methods	176
	6.4.4	Summary	178
6.5	An Evaluation Strategy for Local Features		178
	6.5.1	Evaluation of Local Feature Extractors	178
	6.5.2	Repeatability	178
	6.5.3	1-Precision-Recall	182
6.6	Conclusion		183
References			184

Contributors

Nataša Živić Chair for Data Communications Systems, University of Siegen, Siegen, Germany

Fawad Ahmed Department of Computer Science and Engineering, HITEC University, Taxila-Cantt, Pakistan

Amir Anees Department of Electrical Engineering, HITEC University, Taxila-Cantt, Pakistan

Robin Fay Chair for Data Communications Systems, University of Siegen, Siegen, Germany

Chen Ling School of Film and TV Arts & Technology, Shanghai University, Shanghai, China

Christoph Ruland Chair for Data Communications Systems, University of Siegen, Siegen, Germany

Jinse Shin Chair for Data Communications Systems, University of Siegen, Siegen, Germany

S. Amir Hossein Tabatabaei Chair for Data Communications Systems, University of Siegen, Siegen, Germany

Obaid Ur-Rehman Chair for Data Communications Systems, University of Siegen, Siegen, Germany

Chapter 1
Hash-Based Authentication of Digital Images in Noisy Channels

Fawad Ahmed and Amir Anees

1.1 Introduction

The advancement in digital technology has provided us with a number of software tools that can be used to tamper digital media contents, for example, images [1]. This creates several challenges in case if a digital image is to be used as a legal evidence. Can we use traditional crypto-hashing and digital signatures to meet the integrity and authentication requirements of digital images? Considering an image as a data stream, cryptographic hash functions like Secure Hash Algorithm 1 (SHA1) along with the Rivest Shamir Adleman (RSA) algorithm can be used for integrity verification and authentication [2]. One of the earliest work to adapt this approach was done by Friedman [3]. There are, however, several reasons that actually impede the direct use of cryptographic techniques for solving multimedia security problems. Unlike textual data that is transmitted through a lossless medium, multimedia data like audio, image, and video may be transmitted and stored using a lossy medium to save bandwidth and storage space. Therefore, using traditional cryptographic hash functions for integrity verification and authentication of multimedia content has a problem that a single bit change in the content due to lossy medium will significantly change the hash value.

Multimedia content verification and authentication therefore require techniques that should be resilient to content preserving manipulations like compression, channel noise, etc., and at the same time be fragile enough to detect malicious manipulations. Further, they are expected to provide the same level of security as provided by traditional cryptographic algorithms. In recent years, a number of interesting and novel

---

F. Ahmed (✉)
Department of Computer Science and Engineering,
HITEC University, Taxila-Cantt, Pakistan
e-mail: fawad@hitecuni.edu.pk

A. Anees
Department of Electrical Engineering,
HITEC University, Taxila-Cantt, Pakistan
e-mail: amiranees@yahoo.com

© Springer International Publishing Switzerland 2015
N. Živić (ed.), *Robust Image Authentication in the Presence of Noise*,
DOI 10.1007/978-3-319-13156-6_1

security technologies have been developed for multimedia content authentication [4–6]. In this chapter, we discuss the challenges in designing robust and secure image hash functions that can be used for checking the integrity of a digital image. As an example, a hash-based image authentication scheme presented in [7] is discussed to illustrate the issues of preserving integrity of digital images transmitted over lossy and noisy channels. We now introduce the general aspects of data security and the core issues related to hash-based image authentication in lossy and noisy channels.

1.1.1 Basic Requirements of Data Security

Generally speaking, data security encompasses four important aspects: confidentiality, integrity, authentication, and nonrepudiation [4, 5]. These aspects are briefly described as follows:

1. *Confidentiality*: This feature enables only the authorized users to see the data. Encryption algorithms are employed to achieve data confidentiality.
2. *Integrity*: Integrity means that the received message has not been altered during transmission. Even if a single bit of the received message is changed, the integrity of the message will be lost.
3. *Authentication*: Authentication enables two or more communicating parties to identify each other. If information is transmitted over a channel, the authentication function enables to check its origin, date, integrity of the information, information source, and intended recipient(s).
4. *Nonrepudiation*: Nonrepudiation ensures that in future, a sender cannot deny his/her previous commitments or actions.

1.1.2 Multimedia Security

Before multimedia security issues are discussed, it is important to understand the difference between multimedia data and multimedia content. Multimedia data represents the exact data values in a multimedia bitstream whereas multimedia content refers to the meaning or semantic of multimedia data. For example, if an image is Joint Photographic Experts Group (JPEG) compressed, the value of its pixels might change but the literal meaning, i.e., its content would remain intact as long as compression does not cause severe distortion in the image. Multimedia data like image, video, or audio are exposed to two types of distortions: *malicious* and *non-malicious*. Malicious manipulation means that data values are changed in such a way that content of a multimedia signal is changed. For example, in Fig. 1.1a, an original image (cameraman) is shown while in Fig. 1.1b the tampered version of the cameraman image is shown. Tampering is done by increasing the length of camera's lens. This is an example of malicious tampering. On the other hand, Fig. 1.1c shows the JPEG compressed version of the cameraman image; the JPEG quality factor was kept at 50. The peak-signal-to-noise ratio

Fig. 1.1 Illustration of malicious and non-malicious manipulation in an image. **a** Original cameraman image. **b** Tampered version of the cameraman image. **c** Compressed version of the image shown in Fig. 1.1a, JPEG QF = 50

(PSNR) [8] between this image and its original version shown in Fig. 1.1a is 31 dB. This shows that the value of pixels of the images shown in Fig. 1.1a and c are not the same. However, from a perceptual point of view, the two images are same. This is an example of non-malicious manipulation that changes value of the pixels, but keeps the semantic of the multimedia data intact.

This brings us to a very important point; a multimedia security algorithm has to face a different type of environment as compared to conventional text-based security systems. For example, in text-based security, the transmission channel is lossless, hence there is no issue of non-malicious operation taking place to the data. For example, in a conventional text-based authentication system, if a single bit of data is changed, the authentication system will report a failure in verification. This is not the case with multimedia systems; the authentication algorithm should by-pass any non-malicious distortion that does not alter its basic semantic. On the other hand, any minute tampering that alters the semantic of the multimedia signal should be positively detected. This makes the design of multimedia security systems a very challenging task as sometimes it becomes extremely difficult to draw a boundary that separates the malicious and non-malicious distortions.

1.1.3 Image Hash

A hash provides a compact representation of any data. Primarily hash is used to form digital signature for the purpose of authentication [2, 9] in a lossless channel. Consider the example of a digital image. A simple way to authenticate a digital image is to use a cryptographic hash function like the SHA1 along with public key encryption algorithms such as the RSA [2, 9]. As discussed above, the problem of using cryptographic hash functions for image hashing is that a single bit change in the image will produce a significantly different hash value. In practice, it is common that a digital image may undergo some content preserving manipulations such as compression, enhancement, etc. These operations may not change the visual appearance of an image, however, the cryptographic hash value will be completely different. From this discussion, we note that image hashing requires techniques which should be somewhat resilient to content preserving manipulations while at the same time be fragile enough to detect malicious manipulations. For digital images, the hash function is referred to as a perceptual hash function (PHF).

Image authentication schemes can be classified into two types: watermark-based and hash-based. Watermarking techniques embed an imperceptible signal into a cover work to form a watermarked image. At the receiver's end, the extracted watermark from the watermarked image is used for authentication [10]. In contrast to watermark-based techniques, hash-based (or digital signature-based) techniques use a PHF to extract a set of features from the image to form a compact representation that can be used for authentication [11]. One main disadvantage of a PHF-based scheme for image authentication is that the hash is an extra overhead that needs to be transmitted or stored besides the image. However, hash-based schemes have many useful advantages as compared to watermark-based authentication schemes, for example:

- There is no distortion introduced in the image as no signal is embedded.
- The size of the authentication watermark is limited by the image embedding capacity. This is not the case with image hash as it is a separate entity.
- It is not necessary that the image hash is transmitted along with the image. It can even be transmitted before or after the image in question is transmitted. Similarly, the image and its respective hash can be stored at two different physical locations to provide enhanced security.

As depicted in Fig. 1.2, hash is generated from the image data and transmitted to the receiver through a secure channel. The image itself is assumed to be transmitted through an insecure channel. The image data may get corrupted due to channel

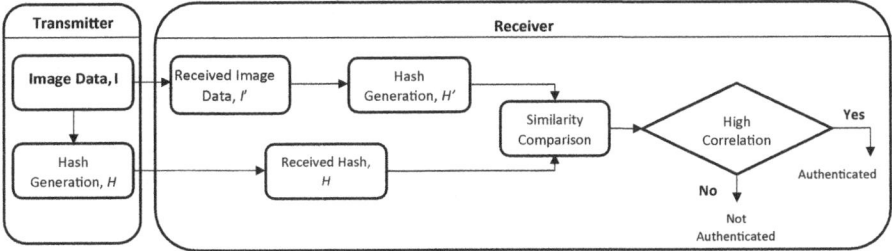

Fig. 1.2 Image hashing framework

noise, its lossy nature, or malicious tampering. At the receiver's end, the image hash is then regenerated from the received image and is compared with the original received hash. A high correlation between the two hashes implies that the received image is positively authenticated. As compared to cryptographic hash functions, the field of image hashing is passing through an evolution stage. Since cryptographic hash functions are matured and well studied for more than a decade, it is very natural to design image hash functions that besides meeting the requirements of multimedia applications follow the security features of cryptographic hash functions. It should be noted that the objectives of a cryptographic hash function and an image hash function are not exactly the same. For example, there is no robustness or tamper localization requirement in case of a cryptographic hash function. However, a good hash function for an image authentication system should address three important issues: robustness to non-malicious manipulations, ability to detect malicious tampering with localization capability and security. Interestingly, all these issues are related to each other. For example, increasing robustness to non-malicious manipulations generally reduces the tamper detection capability and security of the overall scheme. Similarly, if the system is made very sensitive to detect malicious tampering, the robustness parameter will suffer.

1.1.4 Basic Requirements of Image Hashing

Depending on the type of application, a trade-off is usually adopted to balance the three parameters discussed above. The security aspect is of paramount importance in any authentication scheme. A compromise in security means complete failure of the authentication system. It should be noted that the secrecy of a security algorithm does not rest on the obstruction of the algorithm, but rather on the secret key as outlined by Kerckhoff [2]. The secret key, ideally, should allow only the genuine participants to use the algorithm effectively and prohibit all others. Following are some important requirements that are expected from a PHF used to authenticate an image [7, 12–14].

- Robustness against non-malicious manipulations, for example, lossy compression, low-pass/high-pass filtering, minor contrast enhancement, etc.

- The PHF should be highly sensitive to malicious tampering and should be able to detect the location of tampering.
- The PHF should be key dependent. Without knowledge of the correct key, it should be extremely difficult to generate the correct hash.
- A key-based PHF should satisfy the property of modified weak collision [13]. According to this property, given an image, I and its corresponding robust hash, $H(I)$, it should not be feasible to find another image, P such that $H(I) = H(P)$ when I is significantly different from P.
- The PHF should possess a high level of security against counterfeit attacks or attacks that are meant to fool the authentication system.
- By looking at the hash values, it should be extremely difficult for an attacker to guess the contents of the image or reverse engineer the secret key that was used to calculate the hash.

1.2 Review of Image Hashing Schemes

An image hash is generated by selecting relevant features of the image in question. In this section, a number of hashing schemes are reviewed. Friedman [3] in 1993 introduced the idea of a *"trustworthy digital camera"* to authenticate digital images using cryptographic hash functions and public key cryptography. The camera itself is considered as a secure environment which contains an embedded circuitry to generate a digital signature of the captured image. The digital signature is generated by first taking the cryptographic hash (MD5 or SHA1 [9]) of the image and then encrypting the hash with the camera's private key. The receiver can then authenticate the image using the camera's public key. As long as the camera is kept secure, this scheme is extremely secure as it uses standard cryptographic primitives which are now well established. For this scheme to work, it is necessary that the signature verifier should receive exactly the same image that was used while calculating the signature inside the camera. Even an Least Significant Bit (LSB) change in any of the pixel will generate an incorrect digital signature. This is the limitation of using a cryptographic hash function to calculate hash of an image. In a real-world scenario, it is difficult to realize this scheme because digital images often undergo content-preserving operations like compression, format change, etc. This limitation motivated researchers to devise content-based hash functions that are robust to non-malicious manipulations and fragile to malicious manipulations. In general, image hashing schemes can be broadly classified into following categories [11].

1.2.1 Statistic-Based Schemes

Using image statistics such as mean, variance, higher moments, etc., a number of image hashing schemes have been proposed. Xie et al. [15] used mean of an image block to construct short binary representation of an image which they call Approximate Image Message Authentication Codes (AIMAC). They use the Approximate

Message Authentication Code (AMAC) [16] to get the AIMAC. The AMAC is a probabilistic checksum calculated using pseudorandom permutations, masking and majority voting. The image is first divided into nonoverlapping blocks of size 8×8 pixels. The mean of each block is computed and the most significant bit of the mean is taken. In this way, a binary map of the image is formed. The AIMAC bits are then calculated by applying the AMAC algorithm to each row and column of the binary map. The algorithm generates short binary hash of around 100–450 bits and survives JPEG compression to some extent. However, since the AIMAC only uses the Most Significant Bit (MSB) of the block's mean value to generate the hash, therefore, it is quite easy to manipulate an image block in such a way that the visual appearance of the block is changed but the block is still authenticated.

A some what similar scheme having a compact hash is proposed by Lou and Liu [17]. They divide the image into nonoverlapping blocks of 8×8 pixels and calculate the mean of each block. The mean value is then quantized to get the image hash. A 2-bit code is used to get four-level intensity quantization of the mean value of each block. For an image of size 256×256 pixels, the hash size reported is 2048 bits, tolerance to compression up to 0.5 bpp and tamper localization capability. Another novel feature of this scheme is that it can recover the tampered blocks to some extent thus reducing the retransmission data. However, since the scheme uses the mean of each block, which is then further quantized to 2 bits, an attacker can easily manipulate the image block without changing its mean quantization value.

In [18], Lei et al. have proposed an image hashing scheme using the Radon transform. The Random transform is applied on the input image to obtained projections in various orientations and discarding the insignificant coefficients. The moments and central moments were then calculated. Followed by this step, discrete Fourier transform (DFT) was taken on these moments. Normalization and quantization of DFT coefficients was computed to finally generate the image hash.

A similar work to [18] is reported by Tang et al. [19] for color image hashing by using invariant moments. The proposed technique used two different color spaces to generate the hash. The input image was first preprocessed using bicubic interpolation to ensure that all hashes are of equal length. This is followed by Gaussian low-pass filtering to reduce the effect of high-frequency components, noise contamination, etc., on the hash value. The normalized image's Red Green Blue (RGB) color space was transformed into HSI and YCbCr color spaces thus having six different color components. Seven different invariant moments were then calculated for each color component. Thus, the final hash of the input image was generated by concatenating these moments values.

1.2.2 Relation-Based Schemes

One of the early works in designing content-based hash function for image authentication was carried out by Schneider and Chang [20]. They divided the image into blocks and calculated the intensity histogram of each block. To incorporate details of smaller and larger regions of an image, blocks of different sizes were used. The Euclidean distance between intensity histograms was used to measure the distortion in the content of the image in question. There are two serious limitations with this method. First, an image or a specific block can be manipulated in such a way that its intensity histogram remains same or within the distortion bound. The second problem is the storage requirement as the histograms should be stored in the encrypted form otherwise given the knowledge of the block size, an attacker can easily create forged image without even requiring the genuine image.

Lin and Chang [21] have proposed an image authentication technique that relies on the invariant relationship between any two selected Discrete Cosine Transform (DCT) coefficients which are at the same position of two different 8×8 image blocks. To get the hash, the image is first divided into 8×8 pixels nonoverlapping blocks and DCT of each block is taken. Based on a secret key, randomly generated DCT block pairs are obtained. For each DCT block pair, a few frequency bins are randomly selected using the secret key. This also includes the mean value of DCT coefficients in each selected position for all the blocks. The purpose of including the mean value is to prevent an attacker from making uniform changes to all blocks. The hash of each block is generated by comparing the selected frequency bins of that block with the other secretly selected block to get a binary value. Different levels of thresholds are used to record the sign of comparison and absolute difference between the coefficients in increasing precision. The relationship between the selected DCT coefficients remains unchanged for compression/recompression cycle the image has passed through. Due to this reason, this scheme is resilient to JPEG compression while being sensitive towards catching malicious manipulations with localization capability. Radhakrishnan and Memon [12] and Uehara and Safavi-Naini [22] have shown a method to attack this scheme. Through this attack, the secret mapping key can be found.

In [23], Zhao et al. proposed an image hashing scheme based on features obtained from color histogram. The image is first converted from RGB color space to HSI color space as HSI is more suitable to analyze the color perception features then the RGB model. Each component of HSI is then quantified to obtain a one-dimension vector which is further normalized to 24 elements and histogram of these elements is then obtained. The binary sequence obtained from the histogram is permuted using a secret key to obtain the final hash.

1.2.3 Coarse Representation-Based Schemes

Lu and Liao [24] have proposed an image authentication scheme that uses the structure of an image as the digital signature. The image structure is obtained by identifying the parent–child pairs located at the multiple scales in the wavelet domain. Their scheme is based on the observation that the magnitude difference between a parent node and its four child nodes at consecutive scales in the wavelet domain mostly remains preserved for content preserving manipulations like JPEG compression and blurring. However, for malicious manipulations, this relationship changes, which enables the detection of tampering. An important draw back of this scheme is that the signature obtained does not depend on any secret key and can be extracted/verified by anyone who has knowledge of the algorithm. Second, since no secret key is used, therefore for a single image, only one unique signature can be generated.

Monga and Evans [11] used visually significant feature points to generate the image hash. A wavelet-based iterative feature detection algorithm is used to extract these feature points. A feature vector is generated from the extracted feature points. Deterministic and randomized algorithms are developed to extract the hash of the image from the feature vector. The deterministic algorithm does not use any secret key and is weak from the security point of view. This problem is solved using the randomized algorithm that uses a secret key to achieve randomness in various stages of the algorithm. At first the input image is divided into overlapping circular/elliptical regions with randomly selected radii. These regions are then approximated as rectangles using water-filling like approach. The deterministic algorithm is then applied to the random rectangles to get the binary hash value. The proposed scheme is robust to several non-malicious manipulations like JPEG compression, contrast enhancement Gaussian smoothing, scaling, rotation, etc. The scheme detects malicious manipulations like object addition/removal, excessive noise addition, face morphing etc. However, all the malicious attacks reported in [11] are based on changing a significant portion of the image. The feature points are generated mainly in image areas where significant image geometry lies. Due to this reason, feature points do not cover the entire image area, especially the background. It is not very clear how the scheme will perform under small tampering, especially cloning which may not affect the feature points.

Swaminathan et al. [25] have proposed an image hashing scheme that is resilient to geometric and filtering operations. The input image is first preprocessed by applying low-pass filtering, resizing, and histogram equalization. Fourier transform is then applied to the preprocessed image and the resulting transformed image is then converted into polar coordinates to obtain the feature vector. To incorporate security, the feature vector is modulated using pseudorandom numbers generated from a secret key. The result is then quantized and losslessly compressed to obtain the final hash. The normalized Hamming distance between the hashes is used as the performance metric. The results reported in [25] show that the scheme is quite robust to several non-malicious operations like JPEG compression, rotation of around 10°, cropping, shearing. However, it is not clear whether the scheme can detect tampering in a small area of an image.

1.2.4 Matrix-Based Schemes

Monga and Mihcak [26] have used a dimensionality reduction technique called the nonnegative matrix factorization (NMF) to generate the image hash. The NMF technique uses nonnegativity constraints that enable parts-based representation. The results reported in [26] show that the proposed scheme is robust to JPEG compression, rotation, cropping, and resizing. However, tamper detection capability in small image area and security analysis is not reported.

Motivated by the work of Monga and Mihcak [26], Lv and Wang have recently proposed a fast Johnson–Lindenstrauss transform for image hashing [27]. Swaminathan et al. [28] proposed an image hashing scheme resilient to geometric and filtering operations by using the properties of discrete polar Fourier transform. Lu and Hsu [29] have proposed geometric distortion-invariant image hashing scheme by extracting robust meshes from an image. The normalized meshes are then used for generating the hash in the DCT domain. The proposed scheme has good robustness to several types of geometric distortions. This idea is further extended in the watermarking domain [30] in which hash-dependent watermarks are generated using meshes that are extracted from the image to be watermarked. Interestingly, Deng et al. [31] use affine covariant regions for watermark embedding. An affine covariant point detector is used to extract feature points for the constructions of affine covariant elliptical regions. These regions are then normalized into a circle with specific rotation to align with the dominant gradient orientation of the corresponding feature points. The proposed scheme has good robustness to many common image processing operations, especially geometric distortions, like cropping, scale invariance and rotation. The ideas proposed in [30] and [31] can be further investigated to come up with hashing schemes that besides being robust to lossy compression and image filtering can offer good resilience properties for geometric distortions.

1.2.5 Low-Level Feature-Based Schemes

Chan and Chang [32] proposed an authentication scheme for digital images based upon (7,4) Hamming code method. The proposed method consists of three steps. The first step is the embedding procedure in which three parity bits are generated from four MSBs of a pixel using (7,4) Hamming code. The parity bits are then embedded in the same pixel. In case of tampering, both the parity check bits as well the data bits will be destroyed in a tampered area as the parity bits are embedded in the same pixel. To solve this problem, the parity bits of 1 pixel are embedded into another pixel using Torus automorphism. The Torus automorphism takes input of a pixel position whose parity bits are generated and gives the position of that pixel in which parity bits will be embedded. The Torus automorphism has its own drawbacks as it contains the modulus function. To counter this problem, the parity bits are rearranged before they can be embedded into another pixel. The second step is the detecting procedure in which tampered area is first located and then to eliminate the faulty judgments, morphological operations are used. The last step is the recovery procedure that attempts to recover the value of the modified pixels using JPEG-LS pixel predictive scheme. It was assumed that the tampered area does not contain a pixel which has the parity bits of the tampered pixel. So one has the information of parity bits of a pixel whose data bits have to be predicted. The recovering process mainly attempts to predict the MSB of the four data bits of the tampered pixel and then recover the pixel value according to the value of the parity check bits. The experimental results demonstrate that the proposed technique has the capability to resist the impact of noise and can detect the tampered area.

Chan [33] argues that a problem with the technique in [32] is the use of MSBs in the embedding procedure which create a risk of making an incorrect prediction in the recovery step. If there is a mistake in the recovery procedure by making an incorrect prediction, then the corresponding pixel value will be changed significantly. Also, it can affect the prediction of successive pixels as these are dependent on the previous predicted pixel values. To overcome this problem, MSBs are rearranged in a reverse order and then parity bits are generated using (7,4) Hamming code. The technique in [33] also consists of three steps. However, the first step is modified to counter the incorrect prediction by reversing the order of MSBs. The recovering procedure is also modified accordingly to recover the tampered or destroyed pixels. The experimental results reveal that this technique has better capability to tolerate and recover tampered areas as compared to [32].

An interesting idea of creating a hash from ring-based entropy of an image is presented by Tang et al. [34]. The basic aim of this work is to encounter rotation deformation as the rotated and original images have same pixels in a ring. The proposed method has three stages. In the first stage, preprocessing of an input image is done to convert it into a normalized image. The input image is converted into a square image using bilinear transformation to ensure that images with different resolution have same or very similar hashes. Then the color space of the image is converted into YCbCr space. In the second stage, ring division is done by calculating the circle radii and the distance from each pixel to the image center. It is observed that the image contents of the original image's ring were unchanged after rotation. In the final stage, entropy of each ring is calculated to generate the image hash. The length of hash function is equal to the number of rings. Experimental results were compared with existing algorithms with respect to time complexity and length of the hash functions.

1.3 Illustration of an Image Hashing Scheme

From the discussion in the previous section, it is clear that designing an image authentication scheme that offers high robustness and tamper detection capability and is secure like a cryptographic hash function is a challenging task. We now present a wavelet-based image hashing scheme proposed in [7] as an example to illustrate a number of issues that need to be addressed while designing a robust hashing scheme for image authentication. The wavelet transform is used to extract features from an image for hash generation due to its good time–frequency localization property. As pointed out in [7], variations in Low-Low (LL), Low-High (LH), and High-Low (HL) sub-band wavelet coefficients are less for non-malicious distortions like channel noise, JPEG compression, low-pass and high-pass filtering. On the other hand, significant variation is observed in these coefficients for malicious tampering.

An image hash function having good robustness and tamper detection capability but no security cannot be used in a real-life scenario. There are several security considerations that should be taken into account while designing an image hashing scheme for the purpose of image authentication. A number of hash-based image authentication techniques, for example, [14] and [21] use secret key in the feature extraction stage. In such a strategy, an attacker knows the feature space that is used to generate the hash, but cannot identify the exact subset of the features that were used to generate the hash. A security loop hole may exist for images that have a small number of subsets.

An increased level of security can be achieved if the entire feature space that is used to generate the hash is not known to an attacker even if the image to be authenticated and its generated hash is revealed. This can be achieved by using a secret key before the feature extraction stage. By randomly modulating image pixels in the spatial domain, the entire feature space can be made random. In [7], this is referred to as randomized pixel modulation (RPM). The RPM randomly modulates each pixel of the input image using a secret key to obtain the RPM transformed image. The RPM-transformed image is a random pattern which is a function of the input image pixel values and the secret key. The basic idea behind this technique comes from the fact that image features used for hashing depend upon the values of image pixels. Generally speaking, changing the pixel values in a random fashion can possibly make the feature space random. How would it help to provide security? This can be explained using a simple example. Suppose wavelet sub-bands coefficients are directly used to generate the hash. In such a case, an attacker can easily replace an image or its specific portion with some thing else that is perceptually different from the original one but maps to the same hash value. This appears possible because the feature space is known to an attacker. Once the feature space is made random, an attacker cannot predict how the replaced image or its specific portion might contribute towards forming the image hash. A number of properties related to system's security and tamper detection capabilities using the RPM technique are discussed in later sections of this chapter. The image authentication scheme in [7] is now presented in the following sections. There are two major modules: the hash generation module and the image verification module. To distinguish between the parameters of sender and receiver, a bar is used over the receiver's parameters.

1.3.1 Hash Generation Module

Figure 4.2 shows the block diagram of hash generation module. Hash calculation is a two-stage process. First an intermediate hash is calculated which is then permuted to obtain the final hash. Following steps illustrate the hash generation process:

1. Let I be a gray-scale input image of size $M \times M$ pixels whose hash is to be generated. The image I is partitioned into nonoverlapping blocks, each of dimension $G \times G$ pixels. This gives a total of M^2/G^2 blocks. Each block is represented by B_{ki}, where $ki = 0, \ldots, M^2/G^2 - 1$. As reported in [7], the parameter G is

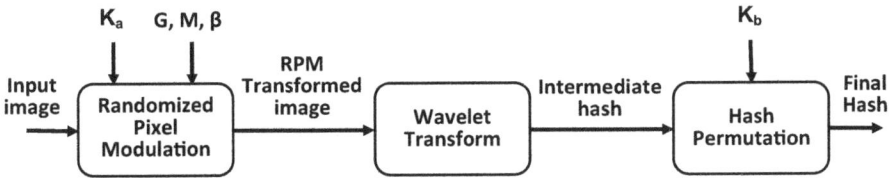

Fig. 1.3 Hash generation module

taken as 16 because it provides a good-trade-off between the tamper detection capability and size of the hash [35].

2. Let $B_{ki}(x, y)$ represent the value of a pixel at the spatial location (x, y) in the block B_{ki}. To enforce security, two secret keys, K_a and K_b are shared between the sender and the receiver (verifier). To make the feature space random and key dependent, pixels in each block are modulated by using a random permutation sequence obtained by the RC4 algorithm [36] that is initialized using the secret key K_a. The RC4 is a well-studied algorithm in cryptography that is capable of generating sequences having a repetition cycle of 10^{100}. Let the permutation sequence of each block be represented by $R_i(m)$, where i is the block index and m is the index of a specific element in R_i. Block by block randomized pixel modulation is independently carried out for each B_{ki} using the permutation sequence R_i to get a new image called the RPM-transformed image, I_{RPM}. Each pixel, $B_{ki}(x, y)^*$ of I_{RPM} is calculated as follows:

$$B_{ki}(x, y)^* = [B_{ki}(x, y) + \beta \times R_i(m)] \quad (1.1)$$

for $0 \leq x, y \leq G - 1$ and $0 \leq m \leq G^2 - 1$.

The range of $R_i(m)$ values is between 1 and 256 to keep a balance between tamper detection capability and randomness in the noisy I_{RPM} image. The parameter β controls the strength of the permutation sequence. Due to randomness in $R_i(m)$ sequence, blocks that appear to be perceptually similar with small difference in pixels values are mapped to different hash values. This point is further illustrated in Sect. 1.6.

3. Random and key-dependent features to generate the image hash are obtained by taking wavelet transform of the I_{RPM} image. A dth level wavelet decomposition of I_{RPM} will yield LL_d, LH_d, HL_d, and HH_d sub-bands, each of dimension $M/(2^d) \times M/(2^d)$. Due to time–frequency localization property of the wavelet transform, each sub-band wavelet coefficient corresponds to their respective spatial area in the I_{RPM} image. This helps to detect tampering with precise localization as tampering in a specific image area would only disturb the corresponding wavelet coefficients, while the remaining wavelet coefficients would remain unchanged.

4. Let u and v point to the position of a wavelet coefficient in a sub-band, where $0 \leq u, v \leq \mathcal{Z}$. The intermediate hash is calculated by the following two equations

$$\mathcal{H}_{\tilde{I}_1}(u, v) = LL_d(u, v) + LH_d(u, v), \quad (1.2)$$

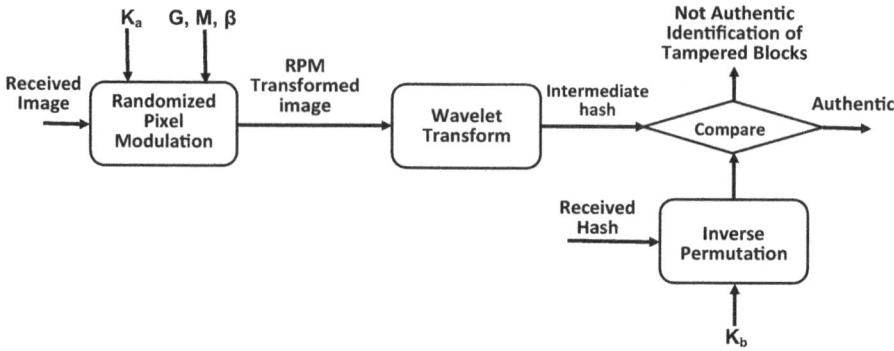

Fig. 1.4 Image verification module

$$\mathcal{H}_{\bar{I}_2}(u, v) = LL_d(u, v) + HL_d(u, v), \quad (1.3)$$

for all $0 \leq u, v \leq \mathcal{Z}$.

5. The entries of $\mathcal{H}_{\bar{I}_1}$ and $\mathcal{H}_{\bar{I}_2}$ are permuted with the secret key, K_b to get $\mathcal{H}_{I_1}$ and $\mathcal{H}_{I_2}$, respectively. The elements of the two matrices $\mathcal{H}_{I_1}$ and $\mathcal{H}_{I_2}$ contains the final hash values for the image I.

1.3.2 Image Verification Module

The module to verify the integrity of the received image is shown in Fig. 1.4. The input to this module is the received image, $\bar{I}$, the received hash matrix $[\mathcal{H}_{I_1} \mathcal{H}_{I_2}]$ calculated at the sender's side and the system parameters M, K_a, K_b, β and G and d. The system parameters are required to be transmitted through a secure channel while the image I may be transmitted through an insecure, noisy, and lossy channel. The steps involved in the image verification are outlined as follows:

1. Using the system parameters, the received image $\bar{I}$ of size $M \times M$ pixels is passed through the same steps as outlined in Sect. 1.3.1 to obtain the dth level wavelet sub-band coefficients.
2. The intermediate hash is calculated using the formula outlined by Eqs. 1.2 and 1.3. Let the intermediate hash of $\bar{I}$ be represented by $\overline{\mathcal{H}}_{\bar{I}_1}$ and $\overline{\mathcal{H}}_{\bar{I}_2}$.
3. Using the secret key K_b, inverse permutation is applied to the received hash matrix $[\mathcal{H}_{I_1} \mathcal{H}_{I_2}]$ to get the matrices of intermediate hash $\mathcal{H}_{\bar{I}_1}$ and $\mathcal{H}_{\bar{I}_2}$ that were calculated at the sender's side.
4. Error matrices E_1 and E_2 are calculated as follows:

$$E_1(u, v) = \left| \mathcal{H}_{\bar{I}_1}(u, v) - \overline{\mathcal{H}}_{\bar{I}_1}(u, v) \right|, \quad (1.4)$$

$$E_2(u, v) = \left| \mathcal{H}_{\bar{I}_2}(u, v) - \overline{\mathcal{H}}_{\bar{I}_2}(u, v) \right|, \quad (1.5)$$

for all $0 \leq u, v \leq \mathcal{Z}$.

5. The maximum error matrix $E_m(u, v)$ is calculated that contains the maximum of the value in either E_1 or E_2.

$$E_m(u, v) = \max(E_1(u.v), E_2(u, v)), \quad (1.6)$$

for all $0 \leq u, v \leq \mathcal{Z}$.

6. Each entry of the matrix E_m is compared with a threshold, η. If an entry of E_m is greater than η, then the corresponding spatial area in $\overline{I}$ shall be considered as tampered and the image $\overline{I}$ will not be positively authenticated.

1.4 Reducing Size of Image Hash Using Quantization

The basic purpose of quantization in image hashing is to reduce the size of a hash. However, interestingly, quantization also enhances the security of the overall hashing system since the final hash is obtained using a cryptographic hash function. To cater for variations in the image due to channel distortion, some extra information is recorded at the transmitter side that is used by the receiver during hash verification. It is later argued that this extra information does not cause any security leak. We shall describe two types of quantization schemes, 3-bit [37] and 4-bit [7], where the number of bits indicates the amount of extra information recorded for a single hash coefficient.

1.4.1 Hash Generation Using 3-Bit Quantization

A 3-bit quantization scheme proposed in [37] is now described. The quantization procedure is depicted in Fig. 1.5. Let ϕ_κ represent a scalar intermediate hash coefficient obtained from the image whose hash is required to be generated and let $\mathcal{Q}_\kappa$ be the quantized value of ϕ_κ. The hash after quantization can be generated using any cryptographic hash function by concatenating all the quantized values. The output of the quantization module consists of two parts. The first part is the cryptographic hash of the input image obtained using any well-known cryptographic hash function. The second part consists of a perturbation vector, λ_κ that is generated as part of the quantization process. Each entry of λ_κ contains 3-bit information which is used at the receiver's end for hash verification.

The information contained in λ_κ is used to adjust the intermediate hash coefficients, $\overline{\phi}_\kappa$ during the image verification stage before performing quantization. This adjustment ensures that if the drift between ϕ_κ and $\overline{\phi}_\kappa$ due to non-malicious operation is less than or equal to the defined threshold η, then $\overline{\phi}_\kappa$ coefficients would quantize to the same value that was calculated in the hash generation phase. To positively authenticate an image, it is necessary that all the coefficients ϕ_κ and $\overline{\phi}_\kappa$ fall in the

Fig. 1.5 3-bit quantization procedure

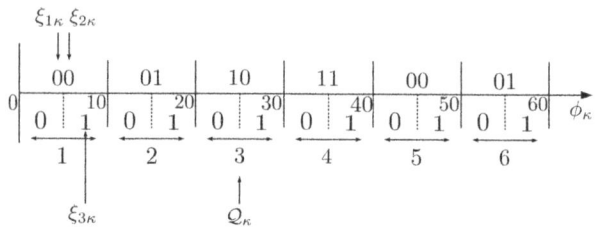

same quantization interval in the hash generation and image verification phases. The quantization procedure is step-wise explained below:

1. Let Q_κ represent a quantization interval, where κ is the index to a specific intermediate hash coefficient. If η is the maximum allowable difference between ϕ_κ and $\overline{\phi}_\kappa$ values due to non-malicious operation, the length of quantization interval will then be equal to η.
2. The sender calculates the quantization intervals for each intermediate hash coefficient ϕ_κ. The quantization interval for ϕ_κ is calculated by dividing ϕ_κ by η and rounding the result using the floor operator.

$$Q_\kappa = \left\lfloor \frac{\phi_\kappa}{\eta} \right\rfloor + 1. \qquad (1.7)$$

3. For each ϕ_κ, a 3-bit value called perturbation information is calculated and recorded in the vector λ_κ. The perturbation information is used at the receiver's end to adjust the value of $\overline{\phi}_\kappa$ coefficients before performing quantization. Let the 3-bit perturbation information for a quantization interval Q_κ be represented by the vector, $[\xi_{1\kappa}\ \xi_{2\kappa}\ \xi_{3\kappa}]$.
4. For a quantization interval Q_κ, the sender calculates $[\xi_{1\kappa}\ \xi_{2\kappa}]$ as follows:

$$[\xi_{1\kappa}\ \xi_{2\kappa}] = (Q_\kappa - 1)\ mod\ 4. \qquad (1.8)$$

As shown in Fig. 1.5, each quantization interval is divided into two equal parts. To show how the bit of $\xi_{3\kappa}$ is recorded, consider the interval between 30 and 40. The parameter $\xi_{3\kappa}$ is recorded as follows:

$$\xi_{3\kappa} = \begin{cases} 0 & : \ 30 \leq \phi_\kappa < 35 \\ 1 & : \ 35 \leq \phi_\kappa < 40. \end{cases} \qquad (1.9)$$

5. All the quantization intervals are then concatenated and hashed using a cryptographic hash function, for example, the SHA1. Let $\mathcal{H}_{SHA1}(\cdot)$ represent the SHA1 hash function and $||$ represents the concatenation operator. The sender calculates the hash, $\mathcal{H}_I$ of the image I using the following equation.

$$\mathcal{H}_I = \mathcal{H}_{SHA1}(Q_0 || Q_1 || Q_2 ||, \ldots, || Q_\kappa). \qquad (1.10)$$

1.4.2 Image Verification Using 3-Bit Quantization

To verify the integrity of the received image, $\bar{I}$, the image verification module would require the corresponding 160-bit hash and the perturbation information calculated during the hash generation phase. Due to non-malicious operation, it is possible that a single or multiple hash coefficients may drift from their original values such that one or more quantization intervals are different from their original counterparts.

According to Eq. 1.10, to positively authenticate an image, it is necessary that $\overline{\mathcal{Q}_\kappa} = \mathcal{Q}_\kappa \forall \kappa$. Therefore, before the receiver calculates $\overline{\mathcal{Q}_\kappa}$, $\forall \kappa \; \overline{\phi_\kappa}$ is adjusted such that $\overline{\mathcal{Q}_\kappa} = \mathcal{Q}_\kappa$ if $|\phi_\kappa - \overline{\phi_\kappa}| \leq \eta$. For a specific $\overline{\phi_\kappa}$, its corresponding recorded vector $[\xi_{1\kappa} \; \xi_{2\kappa} \; \xi_{3\kappa}]$ is used by the receiver to adjust $\overline{\phi_\kappa}$ before calculating $\overline{\mathcal{Q}_\kappa}$. The vector $[\xi_{1\kappa} \; \xi_{2\kappa} \; \xi_{3\kappa}]$ enables the receiver to decide whether shifting in $\overline{\phi_\kappa}$ is required or not. Further, if shifting is required, whether the shifting should be in the positive or the negative direction. Following are the steps to check the integrity of the received image:

1. The receiver will adjust $\overline{\phi_\kappa}$ coefficients to cater for non-malicious operation to get the adjusted coefficients, $\overline{\phi_\kappa^*}$. Let & and | represent logical AND and OR operators, respectively. Let ς_κ be a parameter that enables the receiver to know if $\overline{\phi_\kappa}$ has drifted one quantization interval in either direction. In case if any $\overline{\phi_\kappa}$ coefficient is drifted more than one quantization interval, then it will be considered as a drift due to malicious tampering, hence no adjustment in $\overline{\phi_\kappa}$ will be done. This implies $\overline{\mathcal{Q}_\kappa} \neq \mathcal{Q}_\kappa$. This is the reason that a single quantization interval is made equal to η. The receiver calculates ς_κ as follows:

$$\varsigma_\kappa = \begin{cases} 0 & : \; [\overline{\xi_{1\kappa} \; \xi_{2\kappa}}] = ([\xi_{1\kappa} \; \xi_{2\kappa}] - 1) \bmod 4 \\ 1 & : \; [\overline{\xi_{1\kappa} \; \xi_{2\kappa}}] = ([\xi_{1\kappa} \; \xi_{2\kappa}] + 1) \bmod 4 \\ 2 & : \; \text{otherwise.} \end{cases} \quad (1.11)$$

2. Before calculating $\overline{\mathcal{Q}_\kappa}$, the receiver will adjust $\overline{\phi_\kappa}$ according to the following rule:

$$\overline{\phi_\kappa^*} = \begin{cases} \overline{\phi_\kappa} + \eta & : \; (\varsigma_\kappa = 0) \; \& \; (\xi_{3\kappa} = 0) \\ \overline{\phi_\kappa} + \eta & : \; (\varsigma_\kappa = 0) \; \& \; (\xi_{3\kappa} = 1) \; \& \; (\overline{\xi_{3\kappa}} = 1) \\ \overline{\phi_\kappa} - \eta & : \; (\varsigma_\kappa = 1) \; \& \; (\xi_{3\kappa} = 1) \\ \overline{\phi_\kappa} - \eta & : \; (\varsigma_\kappa = 1) \; \& \; (\xi_{3\kappa} = 0) \; \& \; (\overline{\xi_{3\kappa}} = 0) \\ \overline{\phi_\kappa} & : \; \text{otherwise.} \end{cases} \quad (1.12)$$

3. After obtaining $\overline{\phi_\kappa^*}$ coefficients, the receiver calculates the quantization intervals and the image hash

$$\overline{\mathcal{Q}_\kappa} = \left\lfloor \frac{\overline{\phi_\kappa^*}}{\eta} \right\rfloor, \quad (1.13)$$

$$\mathcal{H}_{\bar{I}} = \mathcal{H}_{SHA1}\left(\overline{\mathcal{Q}_0} || \overline{\mathcal{Q}_1} || \overline{\mathcal{Q}_2} ||, \ldots, || \overline{\mathcal{Q}_\kappa}\right). \quad (1.14)$$

4. The received image $\bar{I}$ will be positively authenticated if:

$$\overline{\mathcal{H}_{\bar{I}}} = \mathcal{H}_I. \tag{1.15}$$

Since the final hash is calculated using the cryptographic hash function, therefore, if tampering in an image area is such that its corresponding $|\phi_\kappa - \overline{\phi_\kappa}| > \eta$ then $\overline{\mathcal{Q}_\kappa} \neq \mathcal{Q}_\kappa$ which implies that $\mathcal{H}_I$ and $\overline{\mathcal{H}_{\bar{I}}}$ will be completely different.

1.4.3 3-Bit Quantization Example

Consider quantization of the hash coefficient $\phi_\kappa = 36$ at the transmitter side. The quantized value and the auxiliary bits obtained using Eqs. 1.7, 1.8, and 1.9 respectively, are, $\mathcal{Q}_\kappa = 4$, $\xi_{1\kappa} = 1, \xi_{2\kappa} = 1$, and $\xi_{3\kappa} = 1$. Now consider the following cases in which at the time of authentication, the value of ϕ_κ is changed due to some distortion.

Case I Assume that the received value of the hash coefficient is changed from 36 to 43, as shown in Fig. 1.6. The auxiliary bits are obtained using Eqs. 1.8 and 1.9 are $\overline{\xi_{1\kappa}} = 0, \overline{\xi_{2\kappa}} = 0$, and $\overline{\xi_{3\kappa}} = 0$. According to Eq. 1.11:

$$[\overline{\xi_{1\kappa}}\ \overline{\xi_{2\kappa}}] = ([\xi_{1\kappa}\ \xi_{2\kappa}] + 1)\,mod\,4$$

$$00 = (11 + 1)\,mod\,4$$

$$0 = (4)\,mod\,4$$

$$0 = 0.$$

Therefore,

$$\varsigma_\kappa = 1.$$

Now by using Eq. 1.12, the modified value of $\overline{\phi_\kappa}$ is obtained as:

$$\overline{\phi_\kappa} = \overline{\phi_\kappa} - \eta,$$
$$\overline{\phi_\kappa} = 43 - 10,$$
$$\overline{\phi_\kappa} = 33.$$

The modified quantized value for the modified $\overline{\phi_\kappa}$ is obtained by using Eq. 1.7, i.e., $\mathcal{Q}_\kappa = 4$, which is correct.

Case II Assume that the received value of the hash coefficient is changed from 36 to 24, as shown in Fig. 1.7. Now the change in the scalar feature is of 12, i.e., $\eta \leq |\phi_\kappa - \overline{\phi_\kappa}| < 1.5\eta$. The auxiliary bits are obtained using Eqs. 1.8 and 1.9 are $\overline{\xi_{1\kappa}} = 1, \overline{\xi_{2\kappa}} = 0$ and $\overline{\xi_{3\kappa}} = 0$. According to Eq. 1.11:

Fig. 1.6 Example of 3-bit quantization

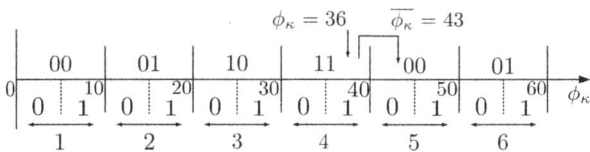

Fig. 1.7 Case II: Value distorted from $\phi_\kappa = 36$ to $\overline{\phi_\kappa} = 24$

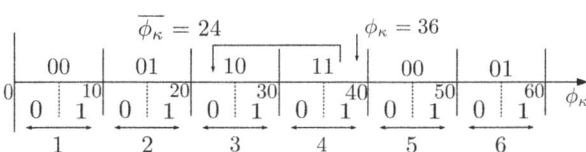

Fig. 1.8 Case III: Value distorted from $\phi_\kappa = 36$ to $\overline{\phi_\kappa} = 49$

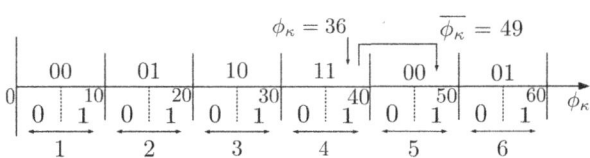

$$\left[\overline{\xi_{1\kappa}}\,\overline{\xi_{2\kappa}}\right] = ([\xi_{1\kappa}\,\xi_{2\kappa}] - 1)\,mod\,4$$
$$10 = (11 - 1)\,mod\,4$$
$$2 = (2)\,mod\,4$$
$$2 = 2.$$

Therefore,

$$\varsigma_\kappa = 0.$$

Now by using Eq. 1.12, the modified value of $\overline{\phi_\kappa}$ is obtained as:

$$\overline{\phi_\kappa} = \overline{\phi_\kappa},$$
$$\overline{\phi_\kappa} = 24.$$

The modified quantized value for the modified $\overline{\phi_\kappa}$ is obtained by using Eq. 1.7, i.e., $\mathcal{Q}_\kappa = 3$, which is wrong.

Case III Now consider a situation in which due to position of a scalar feature, it is possible that a tampering which is $\eta \leq \left|\phi_\kappa - \overline{\phi_\kappa}\right| < 1.5\eta$ would still be corrected. Assume that the received value of the hash coefficient is changed from 36 to 49, as shown in Fig. 1.8. Now the change in the scalar feature is of 13. The auxiliary bits are obtained using Eqs. 1.8 and 1.9 are $\overline{\xi_{1\kappa}} = 0$, $\overline{\xi_{2\kappa}} = 0$, and $\overline{\xi_{3\kappa}} = 1$. According to Eq. 1.11:

$$\left[\overline{\xi_{1\kappa}}\,\overline{\xi_{2\kappa}}\right] = ([\xi_{1\kappa}\,\xi_{2\kappa}] + 1)\,mod\,4$$

Fig. 1.9 Case IV: Value distorted from $\phi_\kappa = 36$ to $\overline{\phi_\kappa} = 51$

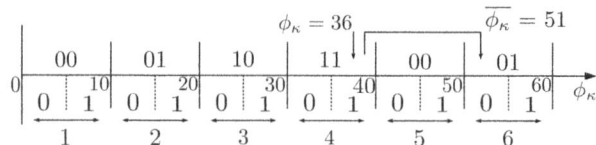

$$00 = (11 + 1) \bmod 4$$
$$0 = (4) \bmod 4$$
$$0 = 0.$$

Therefore,
$$\varsigma_\kappa = 1.$$

Now by using Eq. 1.12, the modified value of $\overline{\phi_\kappa}$ is obtained as:

$$\overline{\phi_\kappa} = \overline{\phi_\kappa} - \eta,$$
$$\overline{\phi_\kappa} = 49 - 10,$$
$$\overline{\phi_\kappa} = 39.$$

The modified quantized value for the modified $\overline{\phi_\kappa}$ is obtained by using Eq. 1.7, i.e., $\mathcal{Q}_\kappa = 4$, which is correct.

Case IV Consider a situation in which if tampering is $|\phi_\kappa - \overline{\phi_\kappa}| \geq 1.5\eta$, then the received scalar feature can not be corrected irrespective of its position. Assume that the received value of the hash coefficient is changed from 36 to 51, as shown in Fig. 1.9. Now the change in the scalar feature is of 15. The auxiliary bits are obtained using Eqs. 1.8 and 1.9 are $\overline{\xi_{1\kappa}} = 0, \overline{\xi_{2\kappa}} = 1$, and $\overline{\xi_{3\kappa}} = 0$. According to Eq. 1.11:

$$\left[\overline{\xi_{1\kappa}}\, \overline{\xi_{2\kappa}}\right] \neq ([\xi_{1\kappa}\, \xi_{2\kappa}] + 1) \bmod 4,$$
$$\text{nor } \left[\overline{\xi_{1\kappa}}\, \overline{\xi_{2\kappa}}\right] \neq ([\xi_{1\kappa}\, \xi_{2\kappa}] - 1) \bmod 4.$$

Therefore,
$$\varsigma_\kappa = 2.$$

By using Eq. 1.12, the modified value of $\overline{\phi_\kappa}$ is obtained as:

$$\overline{\phi_\kappa} = \overline{\phi_\kappa},$$
$$\overline{\phi_\kappa} = 51.$$

The modified quantized value for the modified $\overline{\phi_\kappa}$ is obtained by using Eq. 1.7, i.e., $\mathcal{Q}_\kappa = 6$, which is wrong.

1 Hash-Based Authentication of Digital Images in Noisy Channels

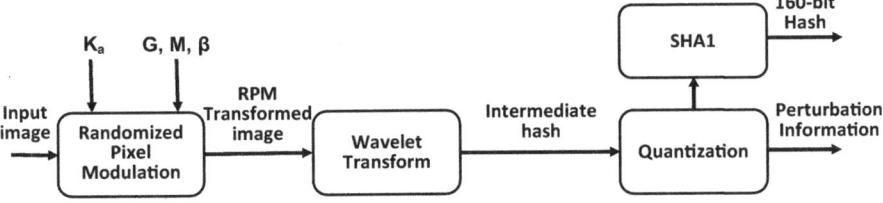

Fig. 1.10 Hash generation module with quantization

1.4.4 A 4-Bit Quantization Scheme

In this section, a 4-bit quantization scheme proposed in [7] is presented to show how the wavelet hash coefficients obtained in Sect. 1.3.1 are quantized. This quantization scheme is a simple extension of the 3-bit quantization method presented in the previous section. The hash generation module shown in Sect. 1.3.1 generates an image hash that consists of permuted wavelet coefficients of the RPM transformed image. The 4-bit quantization scheme takes the intermediate hash of the RPM transformed image and generates a standard cryptographic hash of the input image using any cryptographic hash function. As an example, we have used the SHA1 hash function [9]. Figures 1.10 and 1.12 show the block diagrams of hash generation and image verification modules with the quantization module. The use of a cryptographic hash function not only helps to significantly reduce the size of the hash, but also makes the final hash to appear as a stream of random bits. This improves the security of the overall hashing scheme. In Sect. 1.3.1, the permutation module was used to prevent an attacker from knowing the correspondence between an element of an hash and its respective image block. When quantization is used, the permutation module is not required as the final hash is generated through a cryptographic hash function. In the discussion to follow, the parameters used in the image verification stage are distinguished by putting a line on top of them. For example, ϕ_κ and $\overline{\phi_\kappa}$ are the intermediate hash coefficients at the hash generation and image verification stage, respectively.

1.4.4.1 Hash Generation Using 4-Bit Quantization

In the hash generation phase, the input to the quantization module are the intermediate hash matrices $\mathcal{H}_{\bar{I}_1}$ and $\mathcal{H}_{\bar{I}_2}$ of dimension $\mathcal{Z} \times \mathcal{Z}$. These two matrices are scanned row-wise and concatenated to form a vector, W_κ. Let ϕ_κ represent the intermediate hash coefficients in W_κ, where $0 \leq \kappa \leq 2\mathcal{Z}^2 - 1$. The output of the quantization module consists of two parts. The first part is the 160-bit hash of the input image I obtained using the SHA1 cryptographic hash function. The second part consists of a perturbation vector, λ_κ that is generated as part of the quantization process. The dimensions of W_κ and λ_κ are same and each entry in λ_κ requires 4 bits and corresponds to the entry at the same position in W_κ.

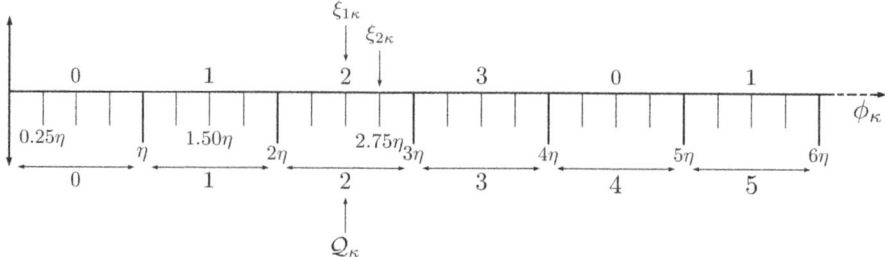

Fig. 1.11 Illustration of quantization procedure

The information contained in λ_κ is used to adjust the intermediate hash coefficients, $\overline{\phi_\kappa}$ in the vector $\overline{W_\kappa}$ during the image verification stage before performing quantization. This adjustment ensures that if the drift between ϕ_κ and $\overline{\phi_\kappa}$ due to non-malicious operation is less than or equal to the defined threshold η, then $\overline{\phi_\kappa}$ coefficients would quantize to the same value that was calculated in the hash generation phase. To positively authenticate an image, it is necessary that all the coefficients ϕ_κ and $\overline{\phi_\kappa}$ fall in the same quantization interval in the hash generation and image verification phases. Figure 1.11 illustrates the concept behind the quantization procedure which is stepwise explained below:

1. Let $\mathcal{Q}_\kappa$ represents a quantization interval, where κ is the index to a specific intermediate hash coefficient in W_κ. If η is the maximum allowable difference between ϕ_κ and $\overline{\phi_\kappa}$ values due to non-malicious operation, the length of quantization interval will then be equal to η.
2. The sender calculates the quantization intervals for each intermediate hash coefficient ϕ_κ in the vector W_κ. The quantization interval for ϕ_κ is calculated by dividing ϕ_κ by η and rounding the result using the floor operator

$$\mathcal{Q}_\kappa = \left\lfloor \frac{\phi_\kappa}{\eta} \right\rfloor. \qquad (1.16)$$

3. For each ϕ_κ, a 4-bit value called perturbation information is calculated and recorded in the vector λ_κ. As described above, the perturbation information is used at the receiver's end to adjust the value of $\overline{\phi_\kappa}$ coefficients in $\overline{W_\kappa}$ before performing quantization. Let the 4-bit perturbation information for a quantization interval $\mathcal{Q}_\kappa$ be represented by two 2-bit variables, $\xi_{1\kappa}$ and $\xi_{2\kappa}$.
4. For a quantization interval $\mathcal{Q}_\kappa$, the sender calculates $\xi_{1\kappa}$ as follows:

$$\xi_{1\kappa} = \mathcal{Q}_\kappa \ mod \ 4. \qquad (1.17)$$

As shown in Fig. 1.11, each quantization interval is divided into four equal parts. To show how the two bits of $\xi_{2\kappa}$ are recorded, consider the interval between η and

2η, i.e., $Q_\kappa = 1$. The 2 bits of the parameter $\xi_{2\kappa}$ are recorded as follows:

$$\xi_{2\kappa} = \begin{cases} 0 & : \quad \eta \leq \phi_\kappa < 1.25\eta \\ 1 & : \quad 1.25\eta \leq \phi_\kappa < 1.50\eta \\ 2 & : \quad 1.50\eta \leq \phi_\kappa < 1.75\eta \\ 3 & : \quad 1.75\eta \leq \phi_\kappa < 2.0\eta. \end{cases} \quad (1.18)$$

5. All the quantization intervals are then concatenated and hashed using the SHA1 hash function. The sender calculates the hash, $\mathcal{H}_I$ of the image I using the following equation:

$$\mathcal{H}_I = \mathcal{H}_{SHA1}\left(Q_0 || Q_1 || Q_2 ||, \ldots, || Q_\kappa\right). \quad (1.19)$$

1.4.4.2 Image Verification Using 4-Bit Quantization

To verify the integrity of the received image, $\overline{I}$, the image verification module would require the corresponding 160-bit hash and the perturbation information calculated during the hash generation phase. The image $\overline{I}$ is first transformed using the RPM technique to get, $\overline{I_{RPM}}$. Wavelet transform of $\overline{I_{RPM}}$ is then taken to get the intermediate hash coefficients, $\overline{\phi_\kappa}$. Due to non-malicious operation, it is possible that a single or multiple hash coefficients may drift from their original values such that one or more quantization intervals are different from their original counterparts (Fig. 1.12).

According to Eq. 1.19, to positively authenticate an image, it is necessary that $\overline{Q_\kappa} = Q_\kappa \forall \kappa$. Therefore, before the receiver calculates $\overline{Q_\kappa}$, $\forall \kappa$ $\overline{\phi_\kappa}$ is adjusted such that $\overline{Q_\kappa} = Q_\kappa$ if $\left| \phi_\kappa - \overline{\phi_\kappa} \right| \leq \eta$. For a specific $\overline{\phi_\kappa}$, its corresponding recorded values $\xi_{1\kappa}$ and $\xi_{2\kappa}$ are used by the receiver to adjust $\overline{\phi_\kappa}$ before calculating $\overline{Q_\kappa}$. The parameters $\xi_{1\kappa}$ and $\xi_{2\kappa}$ enable the receiver to decide whether shifting in $\overline{\phi_\kappa}$ is required or not. Further, if shifting is required, whether the shifting should be in the positive or the negative direction. Following are the steps to check the integrity of the received image:

1. The receiver will adjust $\overline{\phi_\kappa}$ coefficients to cater for non-malicious operation to get the adjusted coefficients, $\overline{\phi_\kappa^*}$. Let ς_κ be a parameter that enables the receiver to know if $\overline{\phi_\kappa}$ has drifted one quantization interval in either direction. In case if any $\overline{\phi_\kappa}$ coefficient is drifted more than one quantization interval, then it will be considered as a drift due to malicious tampering, hence no adjustment in $\overline{\phi_\kappa}$ will be done. This implies $\overline{Q_\kappa} \neq Q_\kappa$. This is the reason that a single quantization interval is made equal to η. The receiver calculates ς_κ as follows:

$$\varsigma_\kappa = \begin{cases} 0 & : \quad \overline{\xi_{1\kappa}} = (\xi_{1\kappa} - 1) \bmod 4 \\ 1 & : \quad \overline{\xi_{1\kappa}} = (\xi_{1\kappa} + 1) \bmod 4 \\ 2 & : \quad otherwise. \end{cases} \quad (1.20)$$

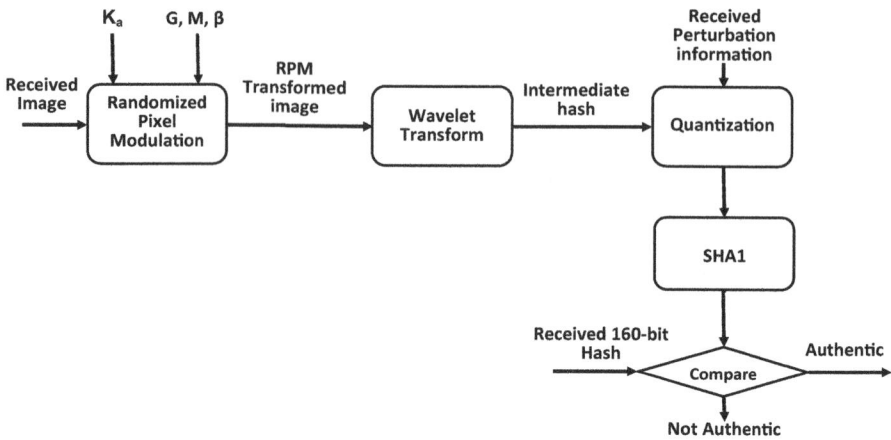

Fig. 1.12 Image verification module with quantization

2. Before calculating $\overline{Q_\kappa}$, the receiver will adjust $\overline{\phi_\kappa}$ according to the following rule:

$$\overline{\phi_\kappa^*} = \begin{cases} \overline{\phi_\kappa} + \eta & : \quad (\varsigma_\kappa = 0) \,\&\, (\xi_{2\kappa} = 0) \\ \overline{\phi_\kappa} + \eta & : \quad (\varsigma_\kappa = 0) \,\&\, (\xi_{2\kappa} = 1) \,\&\, (\overline{\xi_{2\kappa}} = 1 \mid \overline{\xi_{1\kappa}} = 2 \mid \overline{\xi_{1\kappa}} = 3) \\ \overline{\phi_\kappa} + \eta & : \quad (\varsigma_\kappa = 0) \,\&\, (\xi_{2\kappa} = 2) \,\&\, (\overline{\xi_{2\kappa}} = 2 \mid \overline{\xi_{1\kappa}} = 3) \\ \overline{\phi_\kappa} + \eta & : \quad (\varsigma_\kappa = 0) \,\&\, (\xi_{2\kappa} = 3) \,\&\, (\overline{\xi_{2\kappa}} = 3) \\ \overline{\phi_\kappa} - \eta & : \quad (\varsigma_\kappa = 1) \,\&\, (\xi_{2\kappa} = 0) \,\&\, (\overline{\xi_{2\kappa}} = 0) \\ \overline{\phi_\kappa} - \eta & : \quad (\varsigma_\kappa = 1) \,\&\, (\xi_{2\kappa} = 1) \,\&\, (\overline{\xi_{2\kappa}} = 0 \mid \overline{\xi_{2\kappa}} = 1) \\ \overline{\phi_\kappa} - \eta & : \quad (\varsigma_\kappa = 1) \,\&\, (\xi_{2\kappa} = 2) \,\&\, (\overline{\xi_{2\kappa}} = 0 \mid \overline{\xi_{2\kappa}} = 1 \mid \overline{\xi_{2\kappa}} = 2) \\ \overline{\phi_\kappa} - \eta & : \quad (\varsigma_\kappa = 1) \,\&\, (\rho_{2\kappa} = 3) \\ \overline{\phi_\kappa} & : \quad otherwise. \end{cases}$$

(1.21)

3. After obtaining $\overline{\phi_\kappa^*}$ coefficients, the receiver calculates the quantization intervals and the image hash

$$\overline{Q_\kappa} = \left\lfloor \frac{\overline{\phi_\kappa^*}}{\eta} \right\rfloor \tag{1.22}$$

$$\overline{\mathcal{H}_{\bar{I}}} = \mathcal{H}_{SHA1}\left(\overline{Q_0} \| \overline{Q_1} \| \overline{Q_2} \|, \ldots, \| \overline{Q_\kappa}\right). \tag{1.23}$$

4. The received image $\overline{I}$ will be positively authenticated if:

$$\overline{\mathcal{H}_{\bar{I}}} = \mathcal{H}_I. \tag{1.24}$$

Since the final hash is calculated using the cryptographic hash function, therefore, if tampering in an image area is such that $\left|\phi_\kappa - \overline{\phi_\kappa}\right| > \eta$, then $\overline{\mathcal{Q}_\kappa} \neq \mathcal{Q}_\kappa$ which implies that $\mathcal{H}_I$ and $\overline{\mathcal{H}_{\overline{I}}}$ will be completely different.

1.4.4.3 Reduction in Hash Size Due to 4-Bit Quantization

If quantization procedure is not used, the hash of an image would consists of wavelet coefficients in the matrices $\mathcal{H}_{I_1}$ and $\mathcal{H}_{I_2}$. As discussed later in Sect. 1.5.6, after performing experiments on a large database of 256×256 images, it has been observed that each hash coefficient takes 14 bits of storage if a fourth-level wavelet decomposition is used. The total entries in $\mathcal{H}_{I_1}$ and $\mathcal{H}_{I_2}$ for a 256×256 image will be $2 \times 16 \times 16$. With each entry taking 14 bits, the size of the hash will be 896 bytes. If quantization is used, the information contained in λ_κ requires $2 \times 16 \times 16 \times 4$ bits, plus additional 160 bits for the SHA1 hash. Therefore, instead of 896 bytes, the receiver will now require only 276 bytes to perform the authentication. By using the quantization process, the overhead is reduced by 3.24 times thus saving the bandwidth or storage space. An additional benefit of quantization is that it also enhances the security of the system as discussed later in Sect. 1.6.

1.4.4.4 Effect on Robustness and Tamper Detection Due to 4-Bit Quantization

In this section, the effect of quantization on the robustness and tamper detection capability of the system is discussed. As mentioned in Sect. 1.3.2, if any entry of the maximum error matrix E_m is greater than the system threshold η, the corresponding spatial area shall be considered as tampered and the received image $\overline{I}$ will not be positively authenticated. Hence, the threshold η defines a boundary between robustness and tamper detection. The situation is a bit different when quantization is used. The following sections explain the robustness and tamper detection boundaries with numerical examples. The threshold value η is taken as 10.

Effect on Robustness

At the sender's side, assume that the value of a hash coefficient ϕ_κ is 48, for which the quantization bin, $\mathcal{Q}_\kappa = 4$. As a result, the parameters, $\xi_{1\kappa} = 4 \bmod 4 = 0$ and $\xi_{2\kappa} = 3$. Due to channel distortion like noise, compression, etc., suppose the same coefficient at the receiver's side is drifted from 46 to 39, i.e., $\overline{\phi_\kappa} = 39$, therefore the quantization bin calculated at the receiver will be $\overline{\mathcal{Q}_\kappa} = 3$, instead of $\mathcal{Q}_\kappa = 4$. If no adjustment is made, then $\overline{\mathcal{H}_{\overline{I}}} \neq \mathcal{H}_I$, hence $\overline{I}$ will not be positively authenticated. To make adjustment in $\overline{\phi_\kappa}$, the receiver calculates $\overline{\xi_{1\kappa}}, \overline{\xi_{2\kappa}}$, and ς_κ using Eqs. 1.17, 1.18, and 1.20, respectively, yielding $\overline{\xi_{1\kappa}} = 3, \overline{\xi_{2\kappa}} = 3$, and $\varsigma_\kappa = 0$. Using Eq. 1.21, the receiver will adjust $\overline{\phi_\kappa}$ such that $\overline{\phi_\kappa^*} = \overline{\phi_\kappa} + 10 = 49$. Using Eq. 1.22, $\overline{\mathcal{Q}_\kappa} = 4 = \mathcal{Q}_\kappa$. It is easy to check that for any κ, as long as $\left|\phi_\kappa - \overline{\phi_\kappa}\right| \leq \eta, \overline{\mathcal{Q}_\kappa} = \mathcal{Q}_\kappa$ which implies $\overline{\mathcal{H}_{\overline{I}}} = \mathcal{H}_I$.

Condition for Tamper Detection

In light of the above discussion, can we say that tampering will be detected if $|\phi_\kappa - \overline{\phi_\kappa}| > \eta$? Let us consider the case of tampering when $|\phi_\kappa - \overline{\phi_\kappa}| > \eta$, where $\eta = 10$. Assume the same condition at the sender's side, i.e., $\phi_\kappa = 48$, which implies $\xi_{1\kappa} = 0$ and $\xi_{2\kappa} = 3$. At the receiver's side, let $\overline{\phi_\kappa} = 36$, which implies $|\phi_\kappa - \overline{\phi_\kappa}| > 10$. In this case, $\overline{\xi_{1\kappa}} = 3$, $\overline{\xi_{2\kappa}} = 2$, and $\varsigma_\kappa = 0$, therefore, $\overline{\phi_\kappa^*} = \overline{\phi_\kappa} + 10 = 46$. This implies that $\overline{Q_\kappa} = Q_\kappa$, despite the fact that $|\phi_\kappa - \overline{\phi_\kappa}| > \eta$. This is a contradiction from the fact that $\overline{Q_\kappa} \neq Q_\kappa$ if $|\phi_\kappa - \overline{\phi_\kappa}| > \eta$. The reason for this discrepancy is the resolution of the quantization interval. Since each interval is divided into four equal parts, therefore $\overline{\xi_{2\kappa}}$ will not change if the drift in $\overline{\phi_\kappa} < 0.25\eta$. However, it is guaranteed that $\overline{Q_\kappa} \neq Q_\kappa$ if $|\phi_\kappa - \overline{\phi_\kappa}| \geq 1.25\eta$.

Let us take another example in which for the same drift of 11 in $\overline{\phi_\kappa}$, tampering will be detected. At the sender's side, let $\phi_\kappa = 30$, this implies $Q_\kappa = 3$, $\xi_{1\kappa} = 3$, and $\xi_{2\kappa} = 0$. At the receiver's side, let $\overline{\phi_\kappa} = 19$, which implies that $\overline{\xi_{1\kappa}} = 1 \neq (\xi_{1\kappa} - 1) \bmod 4$, $\overline{\xi_{2\kappa}} = 3$ and $\varsigma_\kappa = 2$. There will be no change in $\overline{\phi_\kappa}$, hence, $\overline{\phi_\kappa^*} = \overline{\phi_\kappa}$. Therefore, $\overline{Q_\kappa} = 1 \neq Q_\kappa$. Therefore, $\mathcal{H}_I \neq \overline{\mathcal{H}_I}$. To generalize this discussion, tampering will be detected if $|\phi_\kappa - \overline{\phi_\kappa}| > \eta_\kappa$, where $\eta \leq \eta_\kappa \leq 1.25\eta$. Let x be the distance between any two divisions of a quantization interval, where $0 \leq x \leq 0.25\eta$

$$x_\kappa = \left[\left(\frac{\phi_\kappa}{\eta}\right) \bmod 0.25\right]\eta. \tag{1.25}$$

Depending upon the position of ϕ_κ in a quantization interval, the value of η_κ at which tampering will be detected is given by the following equation:

$$\eta_\kappa = \eta + x_\kappa. \tag{1.26}$$

Hence, in case of quantization, for any κ if $|\phi_\kappa - \overline{\phi_\kappa}| \geq 1.25\eta_\kappa$, it is guaranteed that $\overline{\mathcal{H}_I} \neq \mathcal{H}_I$.

Condition for Detection of Tampered Blocks

If tampering is detected in $\overline{I}$, then $\overline{\mathcal{H}_I} \neq \mathcal{H}_I$. This indicates that $\overline{I}$ is not authentic. The quantization algorithm, however, cannot detect the location of tampering in the received image. Interestingly, the perturbation information stored in λ_κ can be used by the receiver to detect the location of tampering provided $\overline{\xi_{1\kappa}^*} \neq \xi_{1\kappa}$, where $\overline{\xi_{1\kappa}^*} = \overline{Q_\kappa} \bmod 4$. Once the index κ is determined, the corresponding location in the received image $\overline{I}$ can be identified.

1.5 Performance Evaluation

An image hashing scheme is required to be resilient against channel distortions, for example, noise, lossy compression, filtering, contrast enhancement, etc., and should be sensitive enough to detect tampering. It is also expected that the hashing

Fig. 1.13 Image selected to check hash collision. **a** Sailboat image. **b** Car image

scheme should be secure like a cryptographic hash function. In reality, this is very difficult to achieve because introducing robustness in the system effects sensitivity to detect tampering and the level of security that is desired. In this section, a number of parameters that can be used to gauge the effectiveness of an image hashing scheme are discussed. Several experimental results related to these parameters are presented to show the tradeoff between robustness and tamper detection capability of the scheme presented in Sect. 1.3.

1.5.1 The Effect of Beta (β) on Hash Collision

The parameter β in Eq. 1.1 is used for enforcing security by generating a hash that depends not only on the image pixels but also on the secret key. Increasing the value of β increases randomness in the hash values. For the security of an authentication system, it is necessary that collision in the hash values should be as small as possible. Interestingly, the parameter β can be set to control the amount of collision. Increasing β increases randomness in the hash values thus decreasing collision between two wavelet coefficients. We define collision when the difference between two wavelet coefficients corresponding to two different image blocks is less than or equal to the threshold η. With the help of experimental results, the effect of β on hash collision is now presented. Figure 1.13 shows two different images that are used to demonstrate the effectiveness of β on the hash values. Both the images are of size 256×256 pixels. The hash is calculated using fourth-level wavelet decomposition which means that each wavelet hash coefficient corresponds to an image block of size 16×16 pixels. There will be a total of 65,536 comparisons between the two images whose hashes are to be compared. Figure 1.14a shows the collision in the hash values as the parameter β is increased from 0 to 70. When $\beta = 0$, there is no randomized pixel modulation

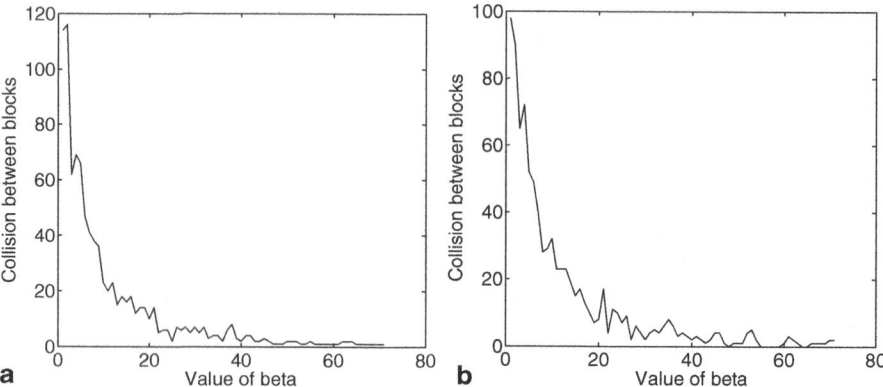

Fig. 1.14 Result of hash collision. **a** Hash collision between sailboat and car image. **b** Hash collision between the sailboat and baboon images

and hence the collision is maximum. Increasing the value of β means making the I_{RPM} image more random. This significantly helps to break the correlation between the original image and its corresponding randomized pixel modulated image. As a result, the hash collision significantly reduces. For the purpose of demonstration, the value of β is taken as 20.

Figure 1.15 demonstrate the effect of randomized pixel modulation. Figure 1.15a shows the original Cameraman image while Fig. 1.15b–d show the RPM-transformed version of the image shown in Fig. 1.15a with $\beta = 1.5$, 20, and 50, respectively. It is evident that increasing β increases the randomness in the RPM-transformed image, thus making the wavelet hash coefficient random.

1.5.2 Robustness to Channel Noise and JPEG Compression

Interestingly, robustness characteristics of an authentication system are not always the same and depend on the type of channel distortion. Further, the texture of an image also has an influence on the amount of change a hash coefficient might undergo due to a specific kind of channel distortion. To demonstrate the robustness against channel noise, the sailboat image is used as a test case and is subjected to zero-mean Gaussian noise with different variance levels. Figure 1.16a shows how the Sailboat image looks like when subjected to zero-mean Gaussian noise having a variance of 0.01. Figure 1.16b shows the maximum deviation in the wavelet hash coefficients, i.e., the difference between the transmitted and received hash values when the received image is subjected to zero-mean Gaussian noise of different levels of noise variance. As the noise level increases, the deviation in the hash coefficient increases. The deviation is around 120 for a variance of 0.01. We further discuss the robustness characteristics of the system against JPEG compression. Figure 1.17a shows the deviation in the hash values when the image to be verified was JPEG compressed with a quality factor

Fig. 1.15 Result of RPM-transformation. **a** Orignal image. **b** RPM-transformed version of the image shown in (**a**): $\beta = 1.5$. **c** RPM-transformed version of the image shown in (**a**): $\beta = 20$. **d** RPM-transformed version of the image shown in (**a**): $\beta = 50$

(QF) of 20. The results are shown for 100 different images in our database. The average deviation in the hash value is around 40. Figure 1.17b shows the deviation in wavelet hash coefficient for the sailboat image when the JPEG QF is changed from 1 to 100. Knowledge about the deviation in the wavelet coefficients helps to determine a threshold value. If the deviation in the hash coefficient is less than the threshold, it would be considered as a non-malicious operation thus authenticating the image positively. On the other hand, a deviation greater than the threshold is considered as malicious tampering.

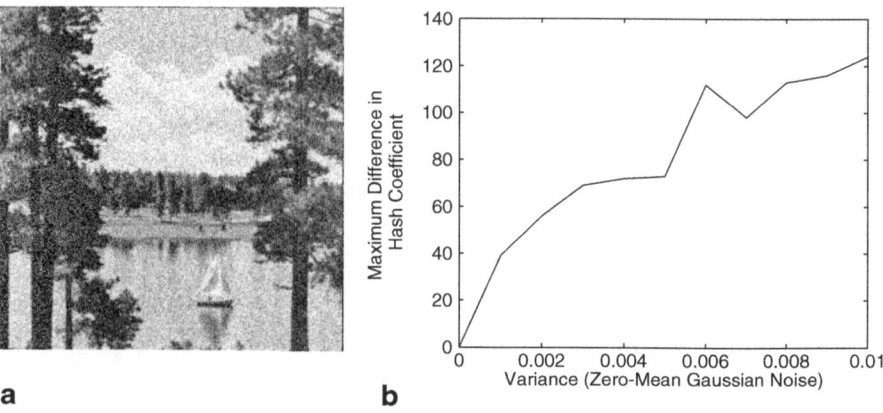

Fig. 1.16 Noisy image and hash coefficient deviation. **a** Sailboat image subjected to zero-mean Gaussian noise (variance = 0.01). **b** Maximum deviation in the hash coefficients due to change in variance

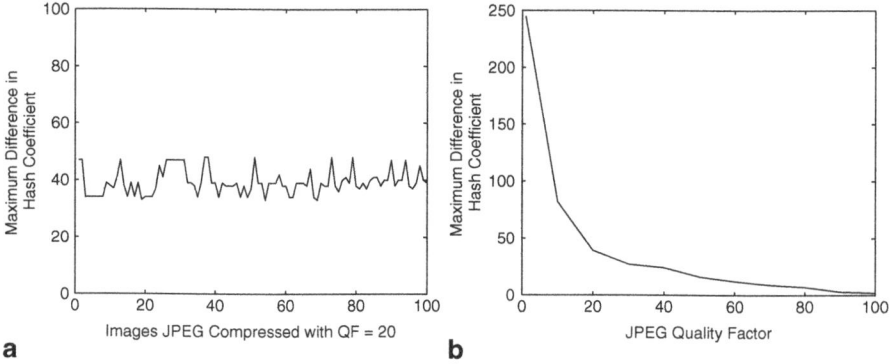

Fig. 1.17 Robustness to JPEG compression. **a** Maximum deviation in the hash coefficients for different images (JPEG QF = 20) **b** Maximum deviation in the hash coefficients for different JPEG quality factors

1.5.3 Threshold Selection

Can we select a single global threshold under all channel distortions? In view of the results shown in Sect. 1.5.2, it appears that defining a global threshold under all channel distortions will not prove efficient. For example, if the receiver has the information that the incoming image would only be subjected to a zero-mean Gaussian noise whose variance can be as large as 0.005, then there is no point in keeping a threshold that can accommodate a variance effect of 0.01. Keeping a higher threshold would make the receiver less sensitive to detect tampering. As another example, consider the case of JPEG compression in which the average deviation in

Fig. 1.18 Tamper detection capability of the hashing scheme. **a** Original bridge image. **b** Tampered version; tampering is shown inside the circle. **c** Result of tamper detection

the hash coefficient is less as compared to noise variance of 0.01. In such a case, a threshold of 60 would suffice. Keeping in view this fact, it is suggested that when a sender is sending an image to a receiver for the purpose of authentication, the sender can estimate the desired threshold in case the channel distortion is known. The threshold parameter can be included in the digital signature. For a specific image, the sender can have some estimate about the authentication system's response to different thresholds. The sender can then select a specific threshold that gives a reasonable trade-off between robustness, tamper detection, and security.

1.5.4 Detection of Tampering

In this section, we show the authentication system capability to detect tampering using the bridge image as an example. Figure 1.18a shows the original bridge image while Fig. 1.18b shows its tampered version with tampered area marked inside a circle. Figure 1.18c shows successful detection of tampering. The deviation in the wavelet coefficient of the tampered area was 142. This result suggests that the system's threshold can be kept around 130–140 if such a tampering is required to be detected. For example, keeping a threshold of 140 means that the system will have the robustness capability to withstand channel noise and JPEG compression to a reasonable extent as evident from the robustness results obtained in Sect. 1.5.2.

This discussion reveals that in robust image hashing, selection of a threshold is a complex issue, as the sender may exactly not know in advance what may be the magnitude of channel distortion and tampering to an image while it will be in transit. However, with the empirical analysis shown above, one has some idea about robustness versus tamper detection capability of the system. As a future research direction, algorithms based on artificial intelligence or other techniques can be devised that can ask the user to specify important image area that need more protection and the type of non-malicious operation that the image may undergo during transmission. Based on this information, the algorithm can give some optimum threshold value that a sender can bound a receiver to use while authenticating the received image.

1.5.5 The Receiver Operating Characteristic Curve

The receiver operating characteristic curve or the ROC curve is a plot of the probability of false positive, P_{FP} versus the probability of false negative, P_{FN} as the system threshold is varied. As mentioned in Sect. 1.3.1, the image hash is formed by using the dth level wavelet coefficients. Due to time–frequency localization of the wavelet transform, each wavelet hash coefficient represents an $\left(M/2^d \times M/2^d\right)$ size spatial image area. With this preamble, the probabilities P_{FP} and P_{FN} are defined as:

$$P_{FP} = \frac{\text{Number of tampered blocks detected as genuine}}{\text{Total number of tampered blocks}}. \quad (1.27)$$

$$P_{FN} = \frac{\text{Number of genuine blocks detected as tampered}}{\text{Total number of genuine blocks}}. \quad (1.28)$$

To estimate P_{FP}, we need to pass a set of tampered blocks through the authentication system and observe whether the system detects them as tampered or genuine as the system's threshold is changed from 0 up to a certain value. It is trivial to note that if the threshold is small, the system will be more sensitive to detect tampered blocks. However, as we increase the threshold, the capability of the system to detect tampered blocks will reduce. Increasing the threshold means adding robustness to the system thus making the system less sensitive to detect tampering. To estimate P_{FN}, a perceptually similar but distorted version of an image is used. Distortion can be made by adding noise, JPEG compression, filtering, etc. The hash between an image and its perceptually similar but distorted version is compared under different thresholds to see how many blocks are detected as tampered as the system's threshold is varied. Ideally for any threshold value, no block should be detected as tampered. However, practically, for very low values of threshold, a slight distortion due to content preserving manipulation like JPEG compression, etc., would cause a genuine image block to be detected as tampered. A plot of P_{FP} versus P_{FN} enables us to know the performance of the authentication system under varying thresholds.

To get an estimate of P_{FP}, an experiment was conducted using the sailboat and baboon images. The sailboat image is considered as a genuine image while the baboon image is considered as a tampered image. The hash comparison is done by comparing each block of the sailboat image with all blocks of the baboon image and identifying deviation in wavelet coefficients which are less than the defined threshold, i.e., $E_m(u,v) < \eta$. For each η, the total number of comparisons are 65,536. A tampered image block shall be considered as genuine if the respective $E_m(u,v)$ for that block is less than η. The P_{FP} is then calculated using Eq. 1.27. To estimate the false-negative probability P_{FN}, JPEG compressed version ($QF = 20$) of the sailboat image is used. The actual hash of the sailboat image is first calculated which is then compared with the hash of the JPEG compressed version of the same image. The comparison though genuine, shall be considered as tampered if for any block, the deviation in the wavelet coefficient is greater than the defined threshold, i.e., $E_m(u,v) > \eta$. The probability P_{FN} is calculated using Eq. 1.28 using the same

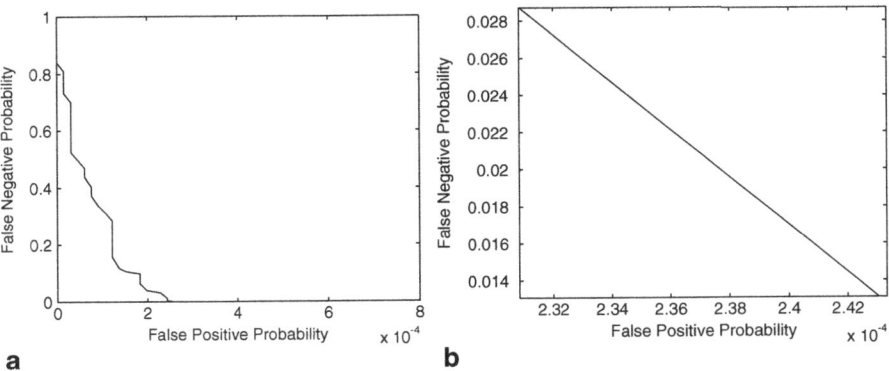

Fig. 1.19 The receiver operating characteristic curve. **a** The ROC curve calculated using Eqs. 1.27 and 1.28. **b** A portion of the ROC curve

values of thresholds that were used to calculate P_{FP}. While estimating the ROC curve, the parameter η was varied from 0 to 100. The result of this experiment is presented in Fig. 1.19a. Figure 1.19b shows an amplified portion of the ROC curve. Even at low values of P_{FP}, the system gives a reasonably low P_{FN}. This is basically a trade-off between robustness, tamper detection capability and security of an authentication system. For example, if P_{FP} is kept at 0.000232, the corresponding P_{FN} is 0.027.

1.5.6 Hash Size

The size of the hash depends upon the size of the image. If quantization is not used, the size of the hash depends upon the size of the wavelet coefficients. Figure 1.20a shows the maximum value of the wavelet coefficients for each image in our database of 100 images. The maximum value of the wavelet coefficient is around 51,000 which require 16 bits of storage. In such a case, the size of the hash for a 256×256 image with a fourth-level wavelet decomposition will be $16 \times 16 \times 2 \times 16 = 8192$ bits or 1024 bytes. The size of the wavelet coefficient depends upon the parameter β as shown in Fig. 1.20b. Since we are using $\beta = 20$, each wavelet coefficient will require 16 bits for storage or transmission.

1.6 Security Analysis

This section presents a number of issues that need to be investigated while analyzing the security of an image hashing scheme. To make the hash secure against attacks, two secret keys K_a and K_b are used in the hash generation and verification stages. The purpose of K_a is to change the gray levels of all the pixels of the original image using the random pixel modulation technique given by Eq. 1.1. For each image pixel, this transformation is random and makes the intermediate hash wavelet coefficients dependent on the input image and the secret key K_a. In addition to this, a permutation stage is also added using the second secret key K_b that is used to randomly permute

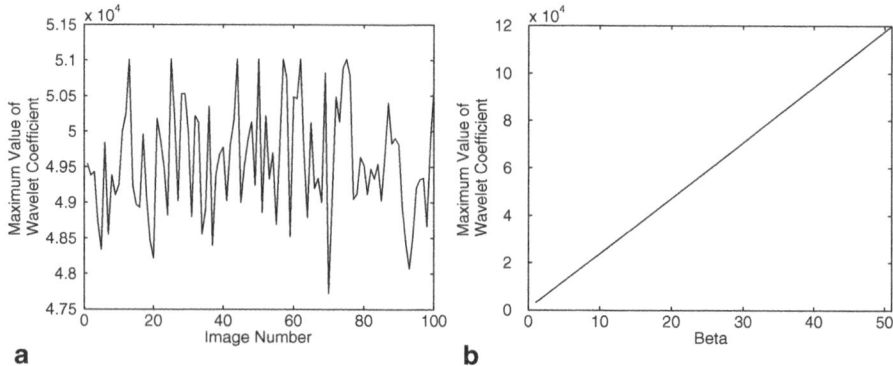

Fig. 1.20 Size of hash. **a** Maximum value of wavelet coefficient for different images. **b** Change in the size of wavelet coefficient as the parameter β changes

the intermediate hash coefficients. By looking at the permuted hash coefficients, an attacker will not be able to relate a hash coefficient to its respective image area. This shall prevent the attacker from launching a brute force attack to estimate K_a. In case if quantization is used, the secret key K_b is not required. This point is discussed in Sect. 1.6.5. In the following subsections, the impact of randomized pixel modulation on the security of the hashing scheme is discussed.

1.6.1 The Impact of Randomized Pixel Modulation on System's Security

A number of image hashing schemes, for example, [14, 21] uses secret key to select the subset of the feature space to generate the image hash. Using this strategy, an attacker does not know which features are selected to form the hash. However, since the original image does not passes through any transformation, the attacker knows the entire feature space. As mentioned in [7], this strategy can have security loop holes that may be exploited by an attacker. For example, the DCT scheme proposed by Sun and Chang [14] forms the hash features by dividing an image into 8×8 nonoverlapping blocks. From each block, the DCT DC coefficient and 3 DCT AC coefficients are selected. The selection of AC coefficients is done using a secret key. To illustrate the issue of feature space in this scheme, consider the 8×8 block marked in the low textured area of the bridge image as shown in Fig. 1.21a. Figure 1.21b shows the DCT coefficients of this block in which most of the DCT AC coefficients have a very small value. If the block marked in Fig. 1.21a is JPEG compressed with a small QF, for example, 60, nearly all the DCT AC coefficients will become zero. This is shown in Fig. 1.21c and is due to the fact that the selected image block belongs to the background with smooth texture. In such a scenario, the impact of the secret key that is used to randomly select the DCT AC coefficients is significantly reduced.

1 Hash-Based Authentication of Digital Images in Noisy Channels

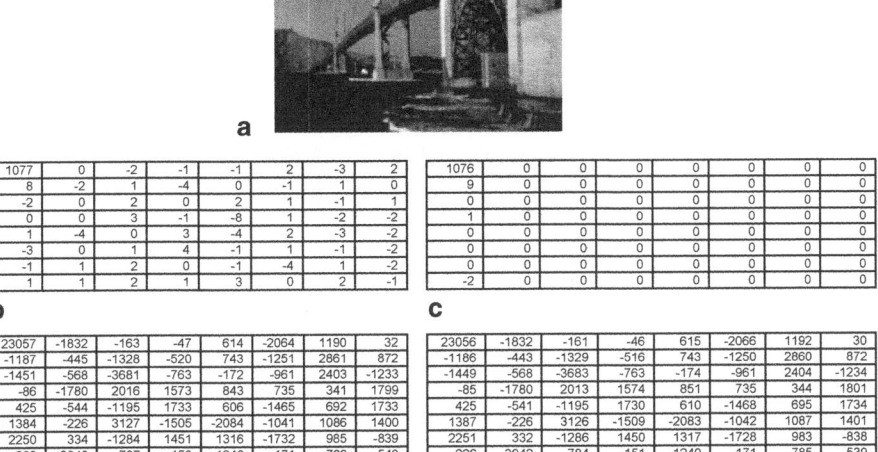

Fig. 1.21 Effect of RPM transformation. **a** an 8 × 8 block marked on the bridge image. **b** DCT coefficient of the 8 × 8 block shown in Fig. 1.21a. **c** DCT coefficient obtained after the 8 × 8 block shown in Fig. 1.21a is JPEG compressed with QF = 60. **d** DCT coefficient of the RPM transformed version of the 8 × 8 block shown in Fig. 1.21a. **e** DCT coefficient obtained after the RPM transformed version of the 8 × 8 block shown in Fig. 1.21a is JPEG compressed with QF = 60

Generally speaking, any hashing scheme that has weak secrecy in the feature selection stage can have such security issues. The scheme presented in this chapter uses a different approach. Instead of using a secret key to randomly select features from a fixed and known feature space, the feature space is first made random using the randomized pixel modulation technique. This enables the RPM transformed image to look like a random noise pattern in which the information of the original image is hidden. The RPM technique further increases the entropy thus making the DCT or other transform domain features both random and available even for an image area that has smooth texture. For example, Fig. 1.21d shows the DCT coefficients for the 8 × 8 image block marked in Fig. 1.21a while Fig. 1.21e shows the DCT coefficients when the same block is JPEG compressed with a $QF = 60$. As evident from these two figures, the DCT coefficients that may be used to form the feature space are random and available even for low texture image area. The same idea applies for the LL, LH, and HL wavelet coefficients that are used in this chapter to generate the hash of an image.

Table 1.1 Statistical analysis of the original and the RPM transformed images for $\beta = 20$

Analysis	Original image	RPM transformed image
Entropy	7.1028	7.9595
Energy	0.1677	0.0191
Contrast	0.4848	7.9178
Homogeneity	0.8956	0.4160
Overall correlation	0.9283	−0.0058
Horizontal correlation	0.9392	0.0085
Vertical correlation	0.9603	0.0072
Diagonal correlation	0.9110	0.0274

1.6.2 Statistical Analysis

To further show the effect of randomized pixel modulation on the original and the RPM transformed images, statistical analysis is performed using correlation, correlation coefficient, entropy, contrast, homogeneity, and energy. Since the hash of the image is formed using the RPM transformed image, hence it is necessary that there should be minimum correlation between this image and the original input image. For the purpose of experiment, the parameter β is taken as 20 and the corresponding results obtained are tabulated in Table 1.1. For example, the entropy of the RPM transformed image is 7.9595, higher than the entropy of original image showing greater randomness. Further, there is less correlation among the image pixels of the RPM transformed image as compared to the original image. Similarly, the contrast value of the RPM transformed image is huge as compared to the value of the original image showing greater variation in image pixels. The correlation coefficient of horizontal, vertical, and diagonal for the original image is shown in Fig. 1.22 and the RPM transformed image in Fig. 1.23, respectively. It is very obvious that the RPM transformation breaks the correlation thus making the hash values random.

Although, we have taken the value of β as 20, increasing the value of β will further improve the randomness in the hash values. This, however, also increases the size of the hash, as discussed later. To show how the parameter β controls the randomness, a graph of the entropy values of the RPM transformed image for different values of β is shown in Fig. 1.24a. It can be seen that by increasing the value of β, the value of entropy also increases thus increasing the randomness. As mentioned above, the RPM transformation significantly helps to make the hash random thus making it extremely difficult for an attacker to make a guess by merely looking at the original image. To validate this phenomenon, the graph of correlation values between an original image hash and its RPM transformed hash is plotted against different values of β from 0 to 100 and is shown in Fig. 1.24b. It can be seen that the correlation between the hash values decreases by increasing the value of β.

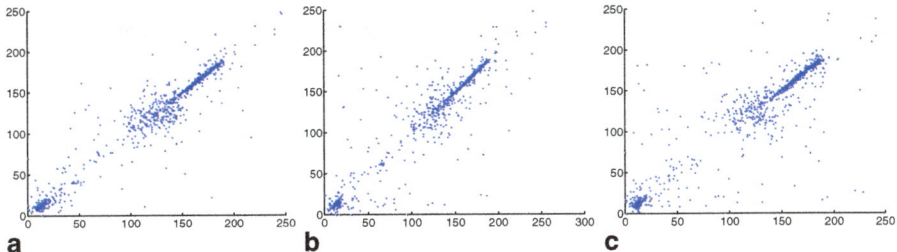

Fig. 1.22 Correlation of the original image. **a** Vertical correlation. **b** Horizontal correlation. **c** Diagonal correlation

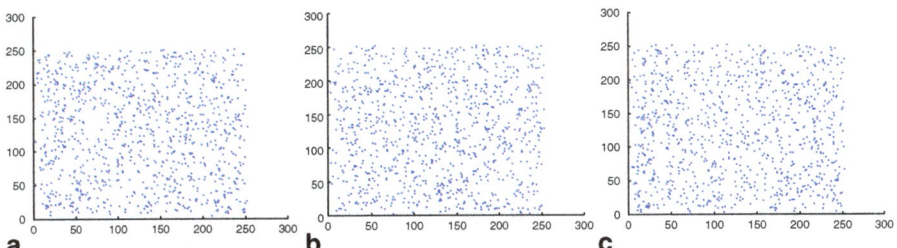

Fig. 1.23 Correlation of the RPM transformed image. **a** Vertical correlation. **b** Horizontal correlation. **c** Diagonal correlation

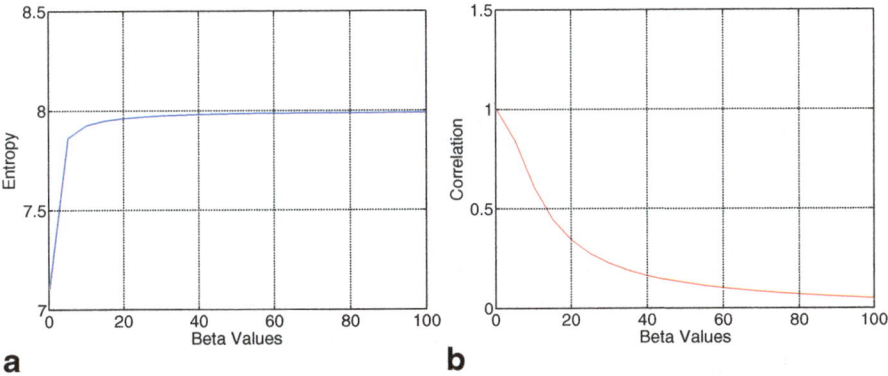

Fig. 1.24 Effect of β on entropy and correlation. **a** Entropy versus β. **b** Correlation versus β

1.6.3 Effect of Secret Key on the Hash

One of the security parameters of an image hashing scheme is the sensitivity of the hash on the secret key. A slightest change in the secret key should significantly change the hash values. As for the hashing scheme under discussion, the secret key is derived using the RC4 algorithm. To observe the effect of the secret key K_a on

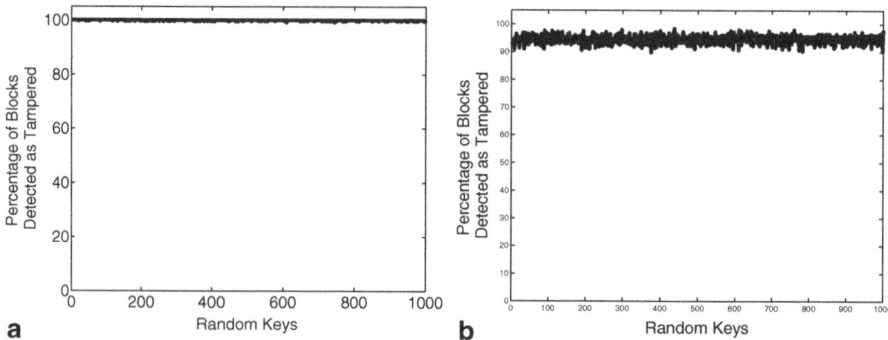

Fig. 1.25 Effect of the change of secret key K_a on the intermediate hash coefficients. **a** Percentage of blocks identified as malicious when $\beta = 20$. **b** Percentage of blocks identified as malicious when $\beta = 1.5$

the intermediate hash coefficients, the hash of the sailboat image was generated and compared with 1000 hashes of the same image generated with 1000 randomly generated keys. The system threshold was kept at 100. Figure 1.25a shows the result of this experiment. Around 99.9 % of image blocks are detected as tampered if the key is changed. This result shows that the RPM technique is very sensitive to change in the secret key and helps to generate image hash which are not only random but also different for different keys. It is important to note that the parameter β plays an important role in key sensitivity. Decreasing β will decrease the key sensitivity as evident from the result shown in Fig. 1.25b.

1.6.4 Probability of Hash Collision

Probability of hash collision means the possibility that a perceptually different image block gets positively authenticated. Due to the RPM transformation, an attacker would not know the feature space and the values of the hash coefficient that were used to form the hash. It will be interesting to estimate the probability of hash collision through an experiment. The sailboat image was used as a test case whose hash was compared with hashes of 100 different images in our database. For each comparison, hash coefficient of every single block of the sailboat image was compared with hash coefficients of all blocks of the target image. A hash collision occurs if for any block of the two images, the difference between the wavelet coefficient is less than the defined threshold, i.e., $E_m(u, v) < \eta$. The images used were of size 256×256 pixels with a fourth-level wavelet decomposition. This means that each wavelet coefficient corresponds to a spatial block of size 16×16 pixels. Hence for each target image, the total number of block comparisons are 65,536. The probability of collision for a single block is calculated by dividing the total number of collisions with the total comparisons. The threshold used was 100, i.e., the system can withstand

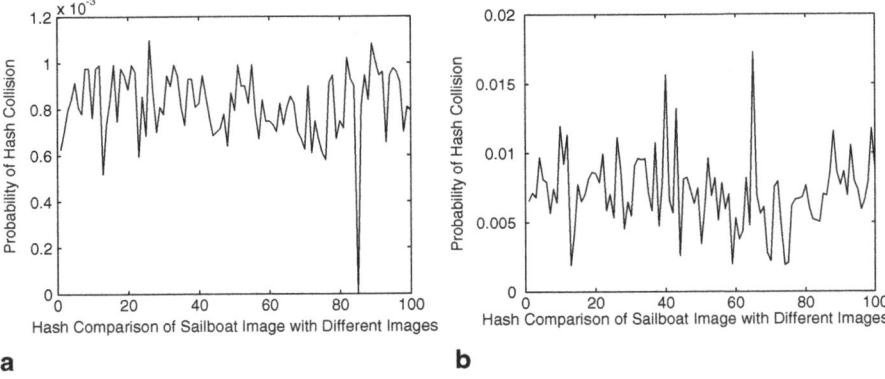

Fig. 1.26 Probability of collision of hash values. **a** Probability of hash collision obtained by comparing the sailboat image with 100 different images for $\beta = 20$. **b** Probability of hash collision obtained by comparing the sailboat image with 100 different images for $\beta = 1.5$

high compression ratios and channel noise. Figure 1.26a shows the result when the parameter β was kept as 20. The average probability of collision for a single block is 0.0008. Decreasing the value of β will increase the probability of collision. For example, when β was taken as 1.5, the average probability of collision for a single block increased from 0.0008 to 0.0075, as shown in Fig. 1.26b. It is to be noted that the probability of hash collision is not that low as we find in cryptographic hash functions. This is because of the robustness feature of an image hash function which is not required in case of a cryptographic hash function.

It should be noted that an image hash function is required to bypass non-malicious operations like compression, etc., which is not the case with cryptographic hash functions. Interestingly, the chances of success of an attacker to randomly replace all blocks of an image with visually different blocks such that all the replaced blocks get positively authenticated is extremely low. If the average probability of collision for a single block is taken as 0.0008, then the probability that all the replaced blocks gets positively authenticated is 0.0008^M, where M in this case is 256. This result explains the importance of the secret key in an image authentication system. If a secret key is not used, the attacker knows the hash coefficient. Therefore, it is easy for an attacker to find collision in the hash values. The attacker will be successful even if the hash is sent encrypted because the hashing algorithm is public.

1.6.5 Quantization and System's Security

As mentioned in Sect. 1.4, the basic purpose of quantization is to reduce the size of the hash. However, the quantization process also helps significantly to enhance the security of the image authentication scheme. When quantization is used, the intermediate hash coefficients are quantized and the quantization intervals are hashed using a cryptographic hash function. The output of the quantization process is the cryptographic hash and the perturbation information. The cryptographic hash is a random string of values and does not reveal any information about the original image, the RPM transformation key, the wavelet hash coefficients or the quantization interval. Similarly, the perturbation information does not reveal the actual quantization interval. Due to randomness in the cryptographic hash values, the secret key K_b is not required when quantization is used. To launch a brute force attack to find the cryptographic hash, an attacker has to search the key space such that all the quantization intervals are correctly found. Even if the attacker is close to finding the key and a single quantization interval differs say by even a unit value, still the hashed output will be entirely different. Hence, an attacker will have no idea whether he/she is even close to the actual key or not. This is the advantage of using a cryptographic hash function because a single bit change in the input significantly changes the hash.

1.7 Conclusion

In this chapter, a wavelet-based image hashing scheme is presented as an example to discuss robustness, security, and tamper detection issues in hash-based image authentication schemes. Unlike other schemes that apply randomness to select the features, a randomized pixel modulation method has been discussed to show how a feature space can be made random. As mentioned in Sect. 1.6, applying randomness to select features does not reveal which features have been selected, however, the feature space used to generate the hash remains open to an attacker. Depending upon the structure of the underlying hashing scheme, this can create security loopholes, especially if entropy in the feature space is not high. In contrast, the scheme presented in this chapter uses RPM to first make the entire feature space random and then extract some useful features to form the image hash. The RPM method is very sensitive to change in the secret key and can effectively detect tampering. Hence by looking at an image, an attacker cannot guess in advance the feature value a specific area of an image would generate. This helps to make it extremely difficult for an attacker to generate counterfeit images that would get positively authenticated.

When image hash is used for authentication, secret key should be used to enforce security. Without a secret key, anyone with the knowledge of the hashing algorithm can generate hash of an image. Once the hash and its corresponding image is known, counterfeit attacks or attacks to fool an authentication system can easily be launched. The scheme described in this chapter uses two secret keys K_a and K_b to randomize the feature space and the sequence of hash values. The system is very sensitive to the change in the secret key. If the final permuted hash is exposed, it will not leak any information that can be used to estimate the secret keys K_a and K_b. In addition, the permuted hash values do not reveal their respective image blocks. In addition, a quantization scheme is also presented to show how the size of an hash can be reduced. The parameters used by the quantization algorithm at the receiver's end do not leak any useful information that may help to estimate the hash or the secret key.

Robustness is an important feature of an image hashing scheme. The scheme under discussion has been shown to be robust to channel noise and JPEG compression that are content preserving operations. Since the image hash is formed by taking into account the LL sub-band wavelet coefficients, the scheme is not robust to other parameters like rotation, scaling, change in brightness, contrast enhancement, etc. This is a performance trade-off as practically it is very difficult to come up with a hashing scheme that is robust to all non-malicious manipulations and still has a high level of security and tamper detection capability. Discarding the LL sub-band coefficients in the hash, for example, will help to improve the system's robustness to contrast enhancement, however, the system would then fail to detect tampering that involves a change in gray values for constant texture images. This is the reason that LL coefficients were used. Combinations of LL, LH, and HL sub-bands provide useful information about image semantics thus exhibiting a better tamper detection capability. The scheme takes the entire image into account for calculating the hash, unlike some of the scheme proposed in the literature that extract features from some specific image area, while leaving the others.

References

1. Li W, Yuan Y, Yu N. Passive detection of doctored JPEG image via block artifact grid extraction. Signal Process; in press.
2. Vanstone SA, Menezes AJ, Oorschot PC. Handbook of applied cryptography. Boca Raton: CRC Press; 1996.
3. Friedman GL. The trustworthy digital camera: restoring credibility to the photographic image. IEEE Trans Consumer Electron. 1993;39(4):905–10.
4. Zeng W, Yu H, Lin C-Y. Multimedia security technologies for digital rights managements. Burlington: Elsevier Inc.; 2006.
5. Furht B, Kirovski D. Multimedia security handbook. Boca Raton: CRC Press; 2005.
6. Lian S. Multimedia content encryption: techniques and applications. Boca Raton: CRC Press; 2008.
7. Ahmed F, Siyal MY, Abbas V-U. A secure and robust hash-based scheme for image authentication. Signal Process. 2010;90(5):1456–70.
8. Cox IJ, Miller ML, Bloom JA. Digital watermarking. San Francisco: Morgan Kaufmann Publishers, Inc.; 2001.
9. Schneier B. Applied cryptography. USA: Wiley; 1996.
10. Wong PW, Memon N. Secret and public key image watermarking schemes for image authentication and ownership verification. IEEE Trans Image Process. 2001;10(10):1593–601.
11. Monga V, Evans BL. Perceptual image hashing via feature points: performance evaluation and tradeoffs. IEEE Trans Image Process. 2006;15(11):3452–65.
12. Radhakrishnan R, Memon N. On the security of the digest function in the SARI image authentication system. IEEE Trans Circuits Syst Video Technol. 2002;12(11):1030–3.
13. Radhakrishnan R, Xiong Z, Memon N. On the security of the visual hash function. Security and Watermarking of Multimedia Contents V. In: Delp EJ III., Wong PW, editors. SPIE vol. 5020, pp. 644–52; 2003.
14. Sun Q, Chang S-F. A robust and secure media signature scheme for JPEG images. J VLSI Signal Process. 2005;41:305–17.
15. Xie L, Arce GR, Graverman RF. Approximate message authentication codes. IEEE Trans Multimed. 2001;3(2):242–52.

16. Gravemen RF, Fu K. Approximate message authentication codes. Proc. 3rd Annual Fedlab Symp. Advanced Telecommunications Information Distribution, vol. 1, College Park, MD; 1999.
17. Lou D-C, Liu J-L. Fault resilient and compression tolerant signature for image authentication. IEEE Trans Consumer Electron. 2000;46(1):31–9.
18. Lei Y, Wang Y, Huang J. Robust image hash in Radon transform domain for authentication. Signal Process: Image Commun. 2011;26:280–8.
19. Tang Z, Dai Y, Zhang X. Perceptual hashing for color images using invariant moments. Appl Math Inf Sci. 2012;6(2S):643S–50S.
20. Schneider M, Chang SF. A content based digital signature for image authentication. Int. Conf. Image Processing, Lausanne, Switzerland; 1996, pp. 227–30.
21. Lin CY, Chang S-F. A robust image authentication method distinguishing JPEG compression from malicious manipulation. IEEE Trans Circuits Syst Video Technol. 2001;11(2):153–68.
22. Uehara T, Safavi-Naini R. On (in) security of a robust image authentication method. In: Y-C Chen et al., editors. LNCS 2532, pp. 1025–32; 2002.
23. Zhao Y, Gu C, Wei W. Image hashing based on color histogram. J Inf Comput Sci. 2012;9(15):4397–404.
24. Lu CS, Liao H-YM. Structural digital signature for image authentication: an incidental distortion resistant scheme. IEEE Trans Multimed. 2003;5(2):161–73.
25. Swaminathan A, Mao Y, Wu M. Image hashing resilient to geometric and filtering operations. IEEE 6th Workshop on Multimedia Signal Processing; 2004, pp. 355–8.
26. Monga V, Mihcak MK. Robust image hashing via non-negative matrix factorizations. IEEE Int. Conf. on Acoustic Speech and Signal Processing; 2006, pp. 225–8.
27. Lv X, Wang ZJ. Fast Johnson-Lindenstrauss transform for Robust and secure image hashing. 10th IEEE Workshop on Multimedia and Signal Processing, Cairns, Australia; 2008, pp. 725–9.
28. Swaminathan A, Mao YM, Wu M. Image hashing resilient to geometric and filtering operations. IEEE 6th Workshop on Multimedia Signal Processing, Italy; 2004. pp. 355–8.
29. Lu C-S, Hsu C-Y. Geometric distortion-resilient image hashing scheme and its applications on copy detection and authentication. Multimed Syst. 2005;11(2):159–73.
30. Lu C-S, Sun S-W, Hsu C-Y, Chang P-C. Media hash-dependent image watermarking resilient against both geometric attacks and estimation attacks based on false positive-oriented detection. IEEE Trans Multimed. 2006;8(4):668–85.
31. Deng C, Gao X, Tao D, Li X. Geometrically invariant watermarking using affine covariant regions. Int. Conf. Image Processing, USA; 2008. pp. 413–6.
32. Chan CS, Chang CC. An efficient image authentication method based on hamming code. Pattern Recognit. 2007;40(2):681–90.
33. Chan CS. An image authentication method by applying Hamming code on rearranged bits. Pattern Recognit Lett. 2011;32:1679–90.
34. Tang Z, Zhang X, Huang L, Dai Y. Robust image hashing using ring-based entropies. Signal Process. 2013;93:2061–9.
35. Coskun B, Memon N. Confusion/Diffusion capabilities of some robust hash functions. Proc. 40th Annual Conf. on Information Sciences and Systems. Princeton, NJ, pp. 1188–93; 2006.
36. Stallings W. Cryptography and network security: principles and practices. Chapter 6. Prentice Hall: Pearson Education, Inc.; 2006. pp. 191–4.
37. Ahmed F, Siyal MY. A robust and secure signature scheme for video authentication. IEEE International Conference on Multimedia and Expo, Beijing; 2007. pp. 2126–9.

Chapter 2
Watermarking for Image Authentication

Chen Ling and Obaid Ur-Rehman

Digital multimedia has become a ubiquitous part of the modern life with the rapid advancement of network communications systems. Digital multimedia content, in the form of images, audios, and videos, is widely used in electronic commerce, national security, forensics, networked communications, social networking websites, and other fields. Through electronic devices, digital multimedia information can be quickly and conveniently shared over the Internet. With the state-of-the-art powerful signal and image processing techniques and the widely available multimedia editing tools with perfect reconstruction capabilities, digital multimedia content is susceptible to malicious manipulations and alterations. It is difficult for users to establish the authenticity of the multimedia content. As a consequence, the copyright owners may suffer huge economic losses and reduced value of the actual multimedia resources. It might also cause a social disorder if digital multimedia content, such as confidential government documents, judicial evidences, or other significant information, were maliciously tampered. Digital images are one of the most widely used types of multimedia and a basis for video. Therefore, the content protection of multimedia, such as images and videos, is becoming an important problem in the study of information security.

Until now, there are two widely used image authentication techniques, i.e., digital signatures and digital watermarking. Digital signatures based authentication is quite mature and already in wide use. However, its shortcoming is that the signature needs extra bandwidth or establishing a separate secure channel for transmission. As a consequence of using secure hash functions, digital signatures are also susceptible to failed authentication due to the avalanche effect. This might happen due to a change of one or more bits of multimedia data, e.g., due to noise, quantization, or

---

C. Ling (✉)
School of Film and TV Arts & Technology, Shanghai University, 200072 Shanghai, China
e-mail: lcrex@shu.edu.cn

O. Ur-Rehman
Chair for Data Communications Systems, University of Siegen, 57076 Siegen, Germany
e-mail: obaid.ur-rehman@uni-siegen.de

compression. Digital watermarking can perform content authentication without the aforementioned shortcomings and they are difficult to be removed or tampered.

2.1 The Basis of Watermarking

In order to cope with the problem of establishing information authenticity, information hiding or watermarking technology has been proposed. Information hiding aims to encode hidden messages in such a way that no one, apart from the sender and the intended recipient, suspects the sole existence of the message. It deals with intelligent ways to embed secret information into unsuspected and rather uninteresting data through clever embedding algorithms. The authenticity of data can be ensured at the receiver based on the embedded secret information. Information hiding makes use of the visual redundancy of the human visual system. It hides digital information into the visual redundancy of other general data; however, the lone existence of this information is invisible to the unintended recipients.

Information hiding can be broadly divided into steganography and digital watermarking. Both steganography and digital watermarking embed data stealthily in noisy signals. Whereas steganography aims for imperceptibility to human senses, digital watermarking considers robustness as its top priority.

Digital watermarking is a covertly embedding technique of digital data with secret information that can be extracted by the recipient. The term "digital watermark" was first coined by Tirkel et al. [1]. The image in which the data to be hidden is embedded, is called the cover image or host. The watermarking process has to be resilient against tampering attacks, keeping the content of a watermark readable in order to be recognizable when extracted by the recipient. Features like robustness and fidelity are the essential features of a watermarking system; however the size of the embedded information has to be considered as well since data becomes less robust as its size increases. Therefore, a trade-off between these features must be considered while developing a watermarking scheme.

Generally, a complete digital watermarking system has three stages: watermark generation, watermark embedding, and watermark extraction for detection and authentication. The general framework is shown in Fig. 2.1.

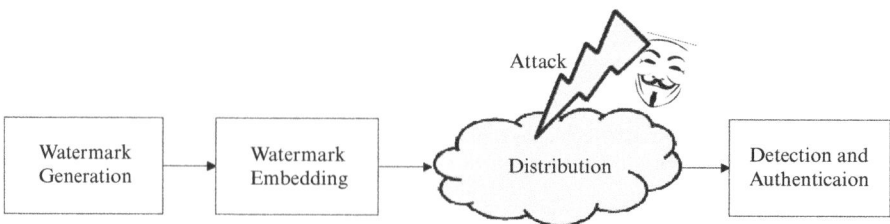

Fig. 2.1 General framework of watermarking

In watermark generation stage, the watermark is created such that its content is unique and complex, making it difficult to be extracted or damaged by the attackers. The image feature f is initially extracted from original cover image I_c.

$$f = \text{Feature}(I_c) \tag{2.1}$$

Feature (·) denotes the feature extraction function, where the feature f uniquely identifies the content of an image. The features can be very simple, such as the grayscale or binary image, edge data of the objects in an image, Wavelet or Fourier coefficients, etc. or they can be much more complicated than these. The image features are then used to generate the watermark, wm, using a secret key K.

$$wm = \text{Generation}(f, K) \tag{2.2}$$

where Generation(·) denotes the watermark generation function.

In cryptography, a secret key is a piece of information (a parameter to the cryptographic algorithm) that determines the functional output of a cryptographic algorithm or a cipher. Without a key, the algorithm would produce no useful results. By changing the key, a significantly different result will be produced. In encryption, a key specifies the particular transformation of plaintext into ciphertext, or vice versa during decryption. The image features generally do not constitute a watermark by itself. Scrambling, random interleaving, and error correcting codes (ECC) are often used for preprocessing to make the secret features more random and useful for watermark generation. The insertion of redundancy (through ECC) in watermarks or in the cover image can improve the extraction process or the reconstruction of watermark after intentional or unintentional distortions. However, ECC can cause collateral effects by increasing the amount of embedded data, which may, in turn, harm the watermark robustness or decrease its data payload, i.e., the capacity of watermark. The most commonly used ECC are: Hamming codes, Bose–Chaudhuri–Hocquenghen (BCH) codes, Reed Solomon (RS) codes, low density parity check (LDPC) codes, and Turbo codes. The watermark is always a pseudo-random sequence or a Gaussian noise sequence.

Embedding of a watermark wm into the cover image image I_c is given by:

$$I_{wm} = \text{E}(I_c, wm) \tag{2.3}$$

where E(·) denotes the watermark embedding function and I_{wm} is the watermarked image. The embedding algorithm should intelligently embed a watermark in the host without destroying the features of the host and making it difficult for an attacker to locate and extract or destroy the embedded watermark.

The transmission of a watermarked image to the receiver is subject to unpredictable distortions and attacks over the communication channel. Some distortions may be unintentional, like channel noise or image compression. Other distortions are intentional, like object removal, replacement, and insertion attacks, etc. Attackers would like to change the original image or remove the watermark to realize their malicious intentions.

At the receiver, the watermarked image I'_{wm} is received, which might have been tampered and therefore different than I_{wm}. A watermark detection and extraction algorithm is usually an inverse process of the watermark embedding algorithm:

$$wm' = D(I'_{wm}) \quad \text{or} \quad wm' = D(I'_{wm}, I_c) \tag{2.4}$$

The watermark wm' is extracted by the extraction function $D(\cdot)$ from the received watermarked image. There are two different types of watermark extraction methods. The first type is called blind watermark extraction, which does not need the original cover image for watermark extraction. The second type needs the cover image to do the same. From the received watermarked image I'_{wm}, the receiver generates a watermark wm'' which might not be the same as wm or wm' due to noise or tampering.

$$f' = \text{Feature}(I'_{wm}) \tag{2.5}$$

$$wm'' = \text{Generation}(f', K) \tag{2.6}$$

By comparing wm' and wm'', it can be determined whether the watermarked image has been tampered or not.

2.2 Classification

There is no uniform criterion for the classification of image watermarking schemes [2]. However, a watermarking system has certain generic requirements which must be met when implemented. According to these requirements, watermarking schemes can be broadly classified into seven categories as shown in Fig. 2.2:

- *Perceptibility*: Watermarking schemes can be classified into visible and invisible watermarking based on the perceptibility criterion.
- *Extraction*: Based on the type of extraction method, watermarking schemes are divided into blind watermarking and non-blind watermarking.
- *Platform*: The watermarking schemes are divided into hardware based and software based according to the system platform used for watermarking.
- *Image compression*: According to the image compression methods, the watermarking schemes can be divided into lossy compression based watermarking and lossless compression based watermarking.
- *Embedding domain*: According to the embedding domain classification, there are spatial domain and transform domain watermarking schemes.
- *Robustness*: According to the robustness criterion, the watermarking schemes can be divided into robust, fragile, and semi-fragile watermarking.
- *Lossless*: According to the quality of the watermarked image, there are irreversible watermarking and reversible watermarking.

2 Watermarking for Image Authentication

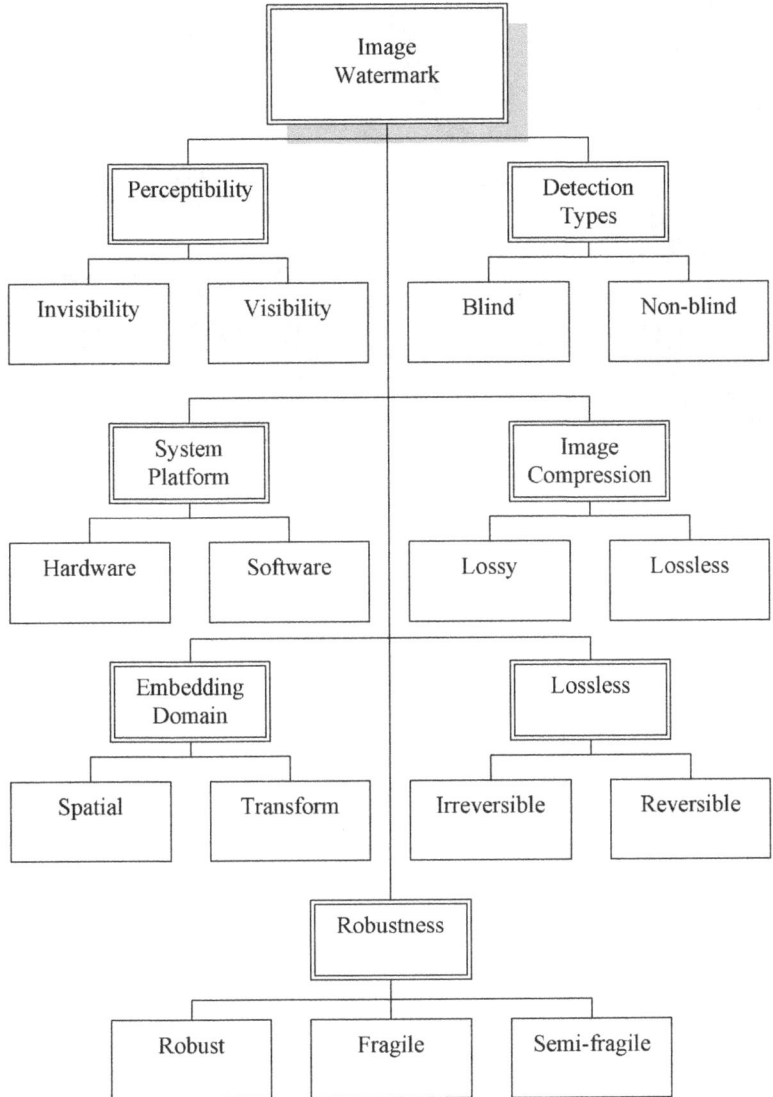

Fig. 2.2 Classification of image watermarking algorithms

2.2.1 Perceptibility

Based on the perceptibility criterion, watermarking schemes are generally classified into visible watermarking and invisible watermarking. The visible watermarking technique is a classical approach, as shown in Fig. 2.3a. The word "WATERMARK" is visible in the original Lena image. Everyone can see the embedded information

Fig. 2.3 Visible watermark (**a**) and invisible watermark (**b**)

that the sender wants others to know. This kind of watermarking appeared in Italy during the thirteenth century. It was used to identify the papermaker or the trading company which manufactured the paper. The watermark shown here was created by a wire sewn onto the paper mold. They are still being used in TV station logo and paper currency, etc.

In invisible watermarking, the original cover image and the watermarked image are indistinguishable, as shown in Fig. 2.3b. The shown image has an embedded watermark in the least significant bit (LSB) of the red components. However, this watermark cannot be detected with a human eye. In this chapter, the term "watermarking" refers to the invisible watermarking.

2.2.2 Detection Types

This classification determines which resources are necessary for an analysis of the extracted watermark from a cover image.

As shown in Eq. 2.4, blind watermarking is a detection type where the original image and watermark data are not available to the receiver. For example, piracy or copy control applications must send different watermarks for each user and the receiver must be able to recognize and interpret these different watermarks.

However, in the non-blind watermarking, the receiver needs either the original image or some information derived from it for the detection process. The same information will also be used in the extraction algorithm. Non-blind watermarking remained a hot research topic in the past few years.

2.2.3 System Platform

According to the system platform, image watermarking can be divided into software-based and hardware-based watermarking. The advantage of software-based watermarking is that it can be easily implemented in the application, not limited by the underlying operating system or hardware. It is also easy to implement any complex algorithm but only limited by the physical computational resources. Generally, software based watermarking is not suitable for real time applications. However, the hardware based watermarking algorithms have relatively powerful processing and computational capabilities. They can meet the requirements of the real-time applications and are especially useful for the limited computing power of small and embedded devices, such as digital cameras.

2.2.4 Image Compression

Images are source coded using compression algorithms to reduce their size for temporal and spatial transmissions. Watermarking can be divided into two types based on the type of image compression schemes: lossless compression based watermarking and lossy compression based watermarking.

Lossless compression is preferred for archival purposes, medical imaging, technical drawings, clip art, comics, etc. Lossy compression methods, especially when used at low bit rates, introduce compression artifacts. Lossy methods are especially suitable for natural images such as photographs in applications where minor loss of fidelity is acceptable to achieve a substantial reduction of the bit rate. There are currently many lossy compression standards in wide use, such as JPEG and JPEG2000.

Watermark can be either directly inserted into the raw image data or integrated during the encoding process after compression. For image compression, the watermarking process can be integrated into image codec or provided as a separate module. The first approach needs to modify the image codec to add watermark embedding and extraction modules. The second one does not change the image codec, but it needs to analyze the encoded stream. It embeds the watermark into the compressed bit stream directly. This is used in the case when the codec cannot be changed and the bit stream can be gained.

2.2.5 Embedding Domain

In the watermark embedding based classification method, two approaches are used for embedding: spatial domain and transform domain. The method used to embed the watermark influences both the robustness against attacks and the detection algorithm, but some methods are very simple and cannot meet the application requirements.

Designing a watermark should consider a trade-off amongst the basic features of robustness, fidelity, and payload.

Spatial domain watermarks insert data in the cover image changing pixels values or image characteristics. The algorithms should carefully weigh the number of changed bits in the pixels against the possibility of the watermark becoming visible. These watermarks have been used for document authentication and tamper detection. The most used spatial domain methods are LSB and spread spectrum (SS).

Transform domain algorithms hide the watermarking data in transform coefficients, thus spreading the data through the frequency spectrum, making it hard to detect. These algorithms are tolerant against many types of signal processing manipulations. The most widely used transforms are discrete cosine transform (DCT) and discrete wavelet transform (DWT).

2.2.5.1 LSB

This is the simplest embedding domain watermarking approach, because the LSBs carry less relevant information and the modification of these bits does not cause perceptible changes. Amongst these approaches, some use only the salient points, by encrypting the watermark before embedding it. In this case, the watermark is embedded in the cover image using a key. The key determines which points will be affected and modified by the embedding process.

A watermark extraction algorithm is the inverse of its embedding algorithm. The least significant pixel bits of an image together with the secret key are used in the decoding algorithm to recover the original watermark. Li et al. [3] proposed a multi-block dependency based fragile watermarking scheme to overcome this shortcoming. The images are split into 8×8 pixel blocks; a 64-bit watermark is generated for each image block which is then equally partitioned into eight parts. Each part of the watermark is embedded into the LSBs of a different image block which is selected by the corresponding secret key.

Figure 2.4 shows the numerical values of Lena's left eye as a block of 16×16 pixel. The cover image and the LSB based watermarked image have only few differences. For example, the value of the first pixel in the cover image is 111, and corresponding pixel has a value of 110 in the watermarked image, which will go unnoticed with a bare human eye.

2.2.5.2 SS

Most of the image watermarking methods are based on the ideas known from SS radio communications, namely additive embedding of a pseudo-noise watermark pattern and watermark recovery by correlation [4].

Figure 2.5 illustrates a simple, straightforward example of SS watermarking. The cover image is the red component of the original RGB color image. The binary watermark bits (represented by $+1$ for a binary 1 and -1 for a binary 0) to be embedded

2 Watermarking for Image Authentication

Cover Image

Watermarked Image

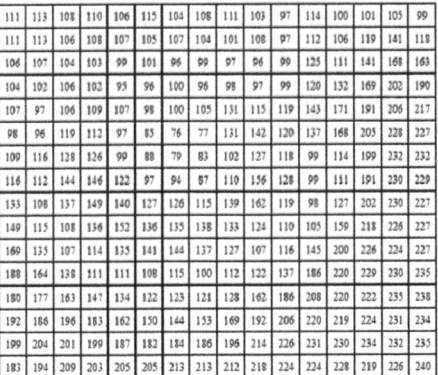

Fig. 2.4 LSB watermarking

are repeated. The spread information bits are then modulated with a cryptographically secure pseudo noise signal, which is scaled according to the visibility criteria and added to the image pixels.

Figure 2.6 illustrates the corresponding watermark detector based on the principle of a correlation receiver. In order to reduce the cross-talk between an image and a watermark, pre-filtering is done to remove low frequencies from the signal, specifically to remove the local mean. If the original cover image is available to the watermark detector, it is advantageous to replace the filtering by subtraction of the original cover image. The filtered watermarked image is then demodulated using exactly the same pseudo-noise signal previously used in watermark embedding. The samples of a correlation signal and a threshold decision yield the output bits. Thus, the result of the watermark decoding is the same watermark information bits that have been embedded.

The SS watermark can also be embedded in the DCT domain. The 1D watermark vector is rearranged into image structure and by transforming the image into the 8×8 DCT domain; the watermark can be added directly to a partially decoded bit stream, making it more robust. However this case is a part of the transform domain algorithm.

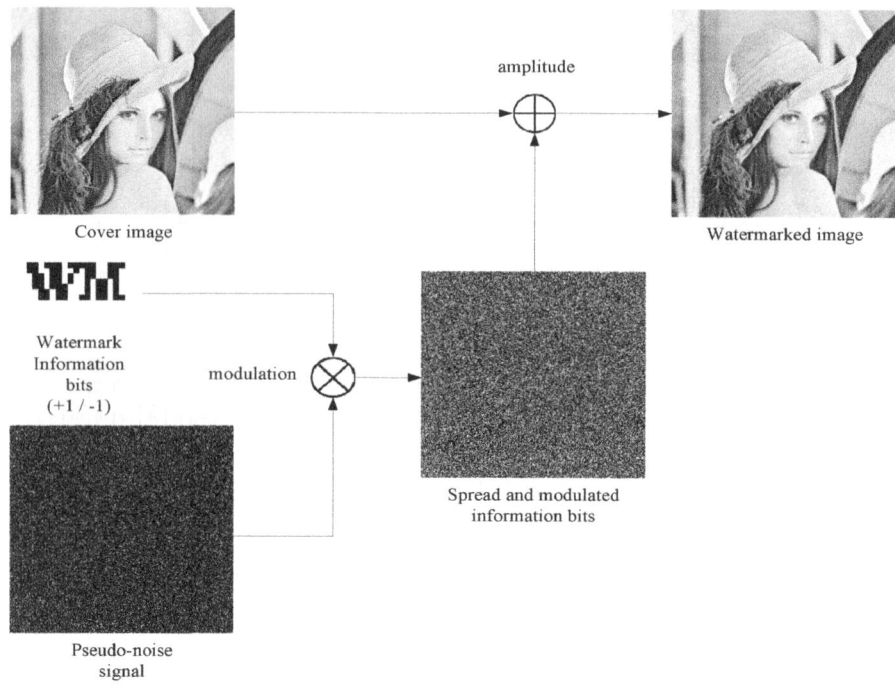

Fig. 2.5 Spread spectrum (SS) watermark embedding

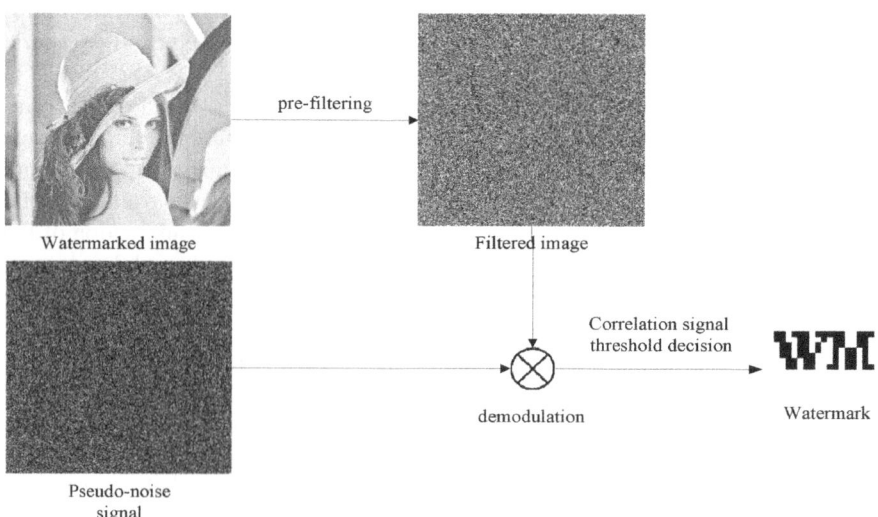

Fig. 2.6 Spread spectrum (SS) watermark extraction

2.2.5.3 DCT

When a watermark is embedded into the cover image spectrum, it does not directly influence the selected image quality. The DCT makes a spectral analysis of the signal and orders the spectral regions from high to low energy. It can be applied globally or in blocks. When applied globally, the transform is applied to all parts of the image, separating the spectral regions according to their energy. When applied to blocks, the process is analogous, only the transform is applied to each block separately.

In particular, DCT is a Fourier-related transform and is similar to the discrete Fourier transform (DFT), but it uses only real numbers. DCTs are equivalent to DFTs but roughly of twice the length, operating on real data with even symmetry (since the Fourier transform of a real and even function is real and even), where in some variants the input and/or output data are shifted by half a sample.

DCT is widely used in image and video compression because it concentrates the high energy components into fewer transform coefficients. The two-dimensional DCT of $N \times N$ blocks are computed and the results are quantized and entropy coded, where N is the block size (typically 8). The result is an 8×8 transform coefficient array in which the (0,0) element (top-left) is the DC (zero-frequency) component and entries with increasing vertical and horizontal index values represent higher vertical and horizontal spatial frequencies.

A typical watermark embedding algorithm based on this approach has the following steps:

1. Segment the cover image into nonoverlapping blocks, say of 8×8 pixels each;
2. Apply forward DCT to each of these blocks;
3. Apply a block selection criterion. For example, the rich texture blocks may be chosen to be embedded into the watermark;

$$F(u,v) = c(u)c(v) \sum_{i=0}^{N-1} \sum_{j=0}^{N-1} f(i,j) \cos\left[\frac{\pi(2i+1)}{2N}u\right] \cos\left[\frac{\pi(2j+1)}{2N}v\right]$$

$$0 \le u \le N-1, 0 \le v \le N-1$$

$$c(u) = \begin{cases} \sqrt{\frac{1}{N}}, u=0 \\ \sqrt{\frac{2}{N}}, u \ne 0 \end{cases}, c(v) = \begin{cases} \sqrt{\frac{1}{N}}, v=0 \\ \sqrt{\frac{2}{N}}, v \ne 0 \end{cases} \quad (2.7)$$

Where N is the block size, e.g., $N=8$.

4. Apply a coefficient selection criterion, e.g., the mid- or high-frequency AC coefficients maybe chosen;
5. Embed watermark by modifying the selected coefficients, e.g., according to the watermark bit 1 or 0, the AC coefficients maybe modified;
6. Apply inverse DCT transform on each block.

$$f(i,j) = \sum_{u=0}^{N-1}\sum_{v=0}^{N-1} c(u)c(v)F(u,v)\cos\left[\frac{\pi(2i+1)}{2N}u\right]\cos\left[\frac{\pi(2j+1)}{2N}v\right]$$

$$c(u) = \begin{cases} \sqrt{\frac{1}{N}}, u = 0 \\ \sqrt{\frac{2}{N}}, u \neq 0 \end{cases}, c(v) = \begin{cases} \sqrt{\frac{1}{N}}, v = 0 \\ \sqrt{\frac{2}{N}}, v \neq 0 \end{cases} \tag{2.8}$$

2.2.5.4 DWT

The wavelet transform decomposes an image into four channels (LL, HL, LH, and HH) with the same bandwidth thus creating a multi-resolution perspective, as illustrated in Fig. 2.7. The advantage of a wavelet transform is to allow a dual analysis taking into account both the frequency and spatial domains.

Wavelets are being widely studied due to their applications in image compression, with the ability that compression resistant watermarks may be produced. Another interesting feature of the DWT is the possibility to select among different types of filter banks, tuning for the desired bandwidth. When the DWT is applied to an image, the resolution is reduced by 2^K, where K is the number of times the transform was applied. For example, Fig. 2.7 uses $K = 2$.

The wavelet based watermark is inserted by substituting the coefficients of the cover image by the watermark's data. This process improves watermark robustness, but depends on the frequency. The low frequency (LL) channel image will damage the cover image if the coefficients change, which in turn challenges the fidelity propriety. However when this region of the spectrum is marked, a robust watermark against compressions like JPEG and JPEG2000 is achieved. When the middle and high frequency (LH and HL) channels are marked, some benefits against noise and several types of filtering show up. Therefore these algorithms tend to be adapted for human visual system to avoid that smaller modifications in the cover image be perceptible.

2.2.6 Robustness

This feature refers to the ability of watermark detection after several image processing operations. Watermarks cannot survive all kinds of attacks; hence attack resilience must be optimized according to the application. For example, to verify data integrity, a correlation between the received image and the signal is carried out when the watermark is extracted. If differences are found, then manipulations must have been occurred. Therefore, the following classification can be made:

- Robust watermarks

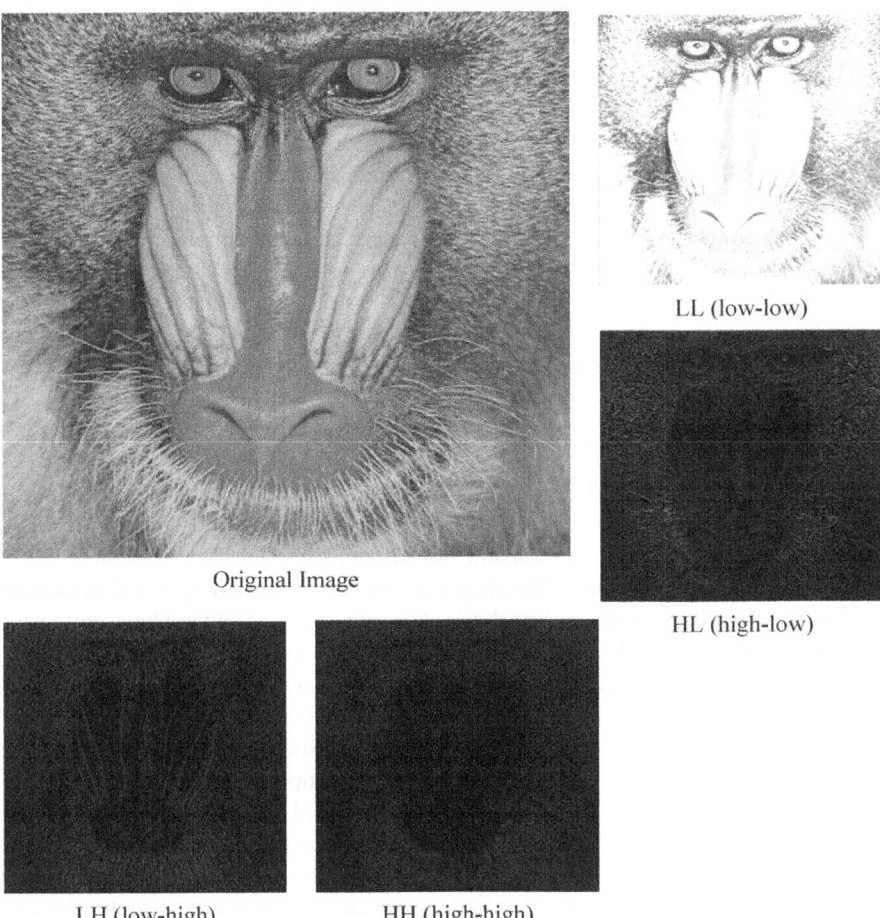

Fig. 2.7 DWT four channels

These watermarks are designed to resist different kind of legal and illegal manipulations. They can be used in a copyright protection, e.g., DVD copy protection. Thabit and Khoo [5] proposed a novel robust reversible watermarking scheme based on using the Slantlet transform matrix to transform small blocks of the original image and hide the watermark by modifying the mean values of the carrier sub-bands. The problem of overflow/underflow has been avoided by using histogram modification.

- Fragile watermarks

These watermarks can be destructed by small manipulations of the watermarked image and therefore they act like digital signatures. Such watermarks have been used for authentication and integrity verification.

The fragile watermark is proposed in order to solve the problem of tampering evidence, namely the information integrity verification. Because digital images are very easy to be altered, the digital camera image does not have a high law significance. Therefore, a trustworthy digital camera concept is proposed, such that when an image is captured, a watermark is embedded in the image. Once the image is manipulated, the image can be authenticated and tampered areas can be located according to the watermark. Thus, the integrity of the picture is protected and the application of digital cameras is enhanced, providing effective technical support for digital rights management. Arsalan et al. [6] proposed a fragile watermarking approach "GA-RevWM" for medical images. GA-RevWM is based on the concept of block-based embedding using genetic algorithm (GA) and integer wavelet transform (IWT). GA based intelligent threshold selection is used to improve the imperceptibility for a fixed payload.

- Semi-fragile watermarking

The semi-fragile watermarking method is a trade-off between robust and fragile watermarking. In case of images, semi-fragile watermarks will be continuously manipulated as a result of compression and conversion. Although these modifications of the image are a result of standard image processing operations, they would be considered as a forgery in case of fragility check. A few bit errors might not have any visual impact on such images due to the nature of such data and human visual perceptual system. Therefore, semi-fragile watermarking is proposed. Semi-fragile watermarking behaves as fragile watermarking against intentional illegal modifications and as robust watermarking against casual legal manipulations like noise. Semi-fragile watermarking is used in image authentication and tamper control. Jiménez-Salinas et al. [7] proposed a novel watermarking algorithm based on the contourlet transform. The embedded watermark is retrieved by computing the correlation between the watermarked coefficients and the pseudorandom sequences. The detection threshold is optimized via Neyman-Pearson criterion; hence the watermark detection can be performed without knowing the value of the watermark strength, which can thus be accurately adapted to each image. The proposed algorithm is robust to common signal processing operations, which could be both intentional and unintentional. Zhang et al. [8] proposed a novel design for image content authentication based on the semi-fragile digital watermarking using side information. It is which combines the RSA public key encryption technique with communication techniques, like SS technology.

2.2.7 Lossless

According to the quality of a watermarked image, watermarking can be classified into reversible and irreversible watermarking. All of the above mentioned watermarking schemes fall in the category of irreversible watermarking, i.e., the watermark embedding is not reversible; causing permanent changes in the original host image after a watermark has been embedded. In applications of high significance, such as military, government, law, and medical images, these changes are hard to be accepted; therefore reversible watermarking was proposed in the patents by Barton [9] and Honsigner et al. [10]. Reversible watermarking method is a technique which enables images to be authenticated and then restored in their original form by removing the digital watermark and replacing the image data that had been overwritten. This makes images acceptable for legal purposes. Reversible watermarking is a new research interest in watermark authentication. This technique for authentication of reconnaissance images is also of interest in military applications. Ni et al. [11] uses the zero or the minimum points of the histogram of an image and slightly modifies the pixel grayscale values to embed data into the image. It can embed more data than many of the existing reversible data hiding algorithms. Tian [12] presented a novel reversible data embedding method for digital images. The differences of neighboring pixel values are calculated, and some difference values are selected for the difference expansion. Poonkuntran and Rajesh [13] proposed a new reversible semi-fragile watermarking scheme for the authentication of digital fundus images that satisfies eight mandatory requirements. The proposed scheme generates a watermark dynamically using chaotic system and it is embedded using integer transform in reversible way. It precisely locates the tampered areas in the images and detects the watermark in a completely blind approach without using the knowledge of both the original image and the watermark. A semi-fragile lossless data hiding scheme based on histogram distribution shift in IWT domain is proposed by Alavianmehr et al. [14]. In the proposed scheme, the transform approximation image is divided into nonoverlapping blocks. In each block, the differences between the neighboring elements are computed and a histogram is made on the difference values. The watermark is embedded into the blocks based on a multi-level shifting mechanism of the histogram. It enables the exact recovery of an original host signal upon extracting the embedded information, if the watermarked image is not affected by any other process.

2.3 Requirements of Watermarking

When a watermarking algorithm is designed and implemented, many requirements should be considered. These requirements are independent of each other and need to be weighed to accomplish the desired purpose, e.g., certain aspects can be decreased to increase the robustness. In this section, some important requirements of a watermarking system are introduced.

2.3.1 Fidelity

This requirement can also be called invisibility. It preserves the similarity between a watermarked image and the original image according to human perception. The watermark must remain invisible notwithstanding the occurrence of small degradations in image brightness or contrast, like the picture in Fig. 2.3b.

As a metric of invisibility, the subjective tests and objective quality measures can be used. Obviously, human eye is the most direct and fastest method to test the fidelity (e.g., "ITU-R BT.500-11" recommendation). The objective quality measure covers many measurements, such as signal to noise ratio (SNR) and peak signal to noise ratio (PSNR).

2.3.2 Capacity

Capacity, also known as the data payload, is the number of message bits that can be inserted into the image through watermarking. Capacity varies with each application. In case of images, a watermark is a static set of bits.

For example, an LSB based watermarking can have a high payload because 1–3 bits of a message can be embedded in every pixel's LSBs for an 8 bit gray image, there is a 64–192 bit capacity just for one 8×8 block without being caught by the human perceptual system. However, the SS method has a low capacity, since it needs many redundant pseudo-random signals to ensure the robustness. DCT and DWT watermarking cannot insert many message bits and are limited by the capacity.

2.3.3 Robustness

The robustness of watermarking means that the watermark can be detected even after the standard image processing operations. The image processing methods include lossy compression, spatial filtering, geometric distortion (rotation, scaling, translation, zooming), etc. Different applications have different requirements for robustness. For fragile watermarking, robustness is just used for complete authentication. Fragile watermarking does not need to be robust against signal processing, but to emphasize sensitive changes. Bit error rate (BER) and normal correlation (NC) are usually used to measure robustness.

2.3.4 Security

The security of watermarking algorithms refers to their ability to resist attacks. During transmission, channel noise might degrade the image quality or damage its content. Such modifications are unintentional or accidental and are inevitable because the common image processing operations or transmission noise introduces these distortions. On the other hand, intentional (or malicious) modifications are attacks on security that use all the available resources to change the image content and at the same time destroy or modify the watermark, making it impossible to be extracted. The methods usually used are signal processing techniques and cryptanalysis. The most common attacks on images are listed below.

Fig. 2.8 a Original airplane image. **b** The airplane serial number "01568" removed on the after-body

2.3.4.1 Malicious Attacks

The most usual malicious attack on images is spatial tampering. In spatial tampering, malicious alterations are performed on the data or on the content of an image. Some of the operations that are used in spatial tampering attack are cropping and replacement, morphing, object addition, modification and removal, etc. These attacks can be efficiently performed with the help of image editing and reconstruction software, such as Photoshop.

Object Removal Attack

In object removal attack, one or more objects in an image are eliminated, e.g., a person may be removed from a photo to destroy the evidence of his or her presence in a scene. This kind of an attack is commonly performed when a person wants to hide something present in the original image. An example is shown in Fig. 2.8, where the serial number "01568" of an airplane, shown in Fig. 2.8a, is removed as shown in Fig. 2.8b.

Object Addition (Insertion) Attack

Object addition or insertion attack involves adding one more objects in an image, whereas they were not present in the original image, e.g., adding something which was not in the original scene. Mostly in the images useful for evidence, an additional object can be inserted into the image, with the help of sophisticated image editing software to mislead the investigation agencies as well as a court. As shown in Fig. 2.9a, an additional airplane is added to the original image.

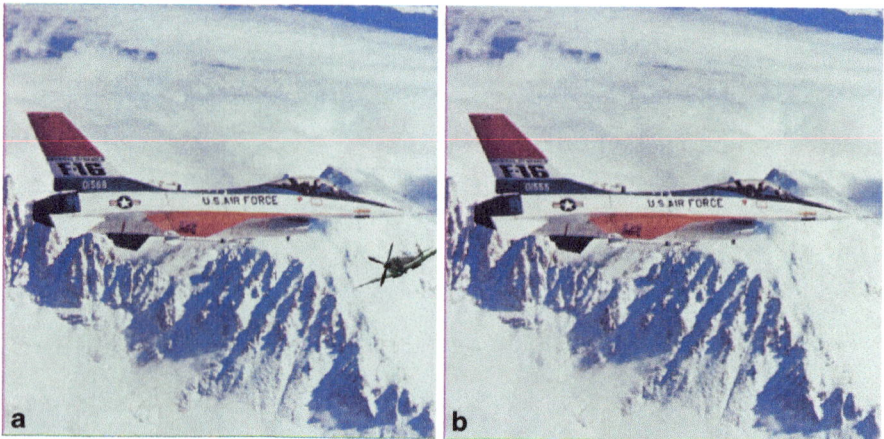

Fig. 2.9 a Another airplane is added in the original image. **b** The airplane serial number "01568" is modified to "01555" on the afterbody

Object Modification Attack

By the object modification attack, an existing object in the image can be modified in such a way that the original identity of that object is lost and a new object appears which is totally different from the original object. Object modification attacks can be done in many forms.

For instance, the size, shape, or color of the existing objects may be changed or the nature of an object and its relation to other objects in an image may be changed with the help of additional manipulations. It is a challenging task, for authentication systems, to detect this kind of attacks since these attacks are performed at the pixel level. The authentication systems should be robust enough to differentiate this kind of attacks from the normal image processing operations, such as compression, etc. Figure 2.9b shows a case that the airplane serial number "01568" is modified to "01555."

Collusion Attack

Another type of malicious attack on images is the collusion attack. Many adversarial attackers share their watermarked images to generate an illegal content (e.g., an image without watermarks).

Suppose f is a cover image and wm is the embedded watermark. The ith watermark wm^i can be seen as the noise to be added to the cover f. This means that the noise watermark is smaller than the cover image f.

If adversaries collected many different watermarked images, they can average these watermarked images to gain a version without watermark, as illustrated in Eq. 2.10.

$$f_{wm} = f + wm^i \tag{2.9}$$

$$\frac{1}{N}\sum_{i=0}^{N} f_{wm}^i = \frac{1}{N}\left(\sum_{i=0}^{N} f + \sum_{i=0}^{N} wm^i\right) \approx \frac{1}{N}\sum_{i=0}^{N} f = f \tag{2.10}$$

2.3.4.2 Incidental Attack

Incidental attacks change the data of the cover image without attempting to target the watermark location. They are non-malicious attacks and include noise addition and image processing without any malicious intent from an adversary. When a watermarked image is transmitted, distortions may be introduced, e.g., as a prevalent noise such as additive white Gaussian noise (AWGN). Two of the incidental attacks or modifications are discussed as follows.

Image Compression

Image compression is the reduction of data size of an image or the number of bits required to represent the image. Applications of compression are in TV broadcasting, remote sensing via satellite, military communication via aircraft, radar, teleconferencing, facsimile transmission of educational and business documents, medical images that arise in computer tomography, magnetic resonance imaging and digital radiology, motion, pictures, satellite images, weather maps, geological surveys, etc. Image compression changes the data of the image (the actual bits) but the content remains unchanged.

Image Enhancement

Image enhancements refer to the accentuation or sharpening of image features, such as boundaries or contrast, to make a graphic display more useful for visualization and analysis. It includes gray level and contrast manipulation, noise reduction, edge sharpening, filtering, interpolation and magnification, pseudo coloring, etc. Like image compression, image enhancement methods also do not change the inherent information content in an image.

2.3.4.3 Secret Key

In cryptography, Kerckhoffs's principle states that a cryptosystem should be secure even if everything about the system, except the key, is publically known. This means that the security of an image authentication system should closely rely on the secret key. Unfortunately, less attention has been paid to openness and transparency of watermarking algorithms. In other words, most authentication methods have an assumption that the attackers do not know that the image has been watermarked. For a good watermarking system, the embedding and extraction algorithms should be public. An adversary will be able to find the watermarked image, extract the original watermark, tamper the image and finally re-embed the extracted original watermark into the tampered image. Therefore, the secret key plays an important role in such a watermarking system.

2.3.5 Tamper Detection, Location, and Recovery

Fragile watermarking is generally used for image content authentication watermarking algorithms. Once the image is attacked, its watermark will be destroyed. Thus, though the attack on image integrity is detectable but due to the traditional fragile watermarking algorithms, based on digital signatures using a hash function, the image authentication will fail due to avalanche effect. Therefore, semi-fragile approach is proposed to solve this problem.

Figure 2.10 shows an example of watermarking. In Fig. 2.10a, 2.10b, the watermark is invisible, whereas a white ball is added and the mouth is removed in Fig. 2.10c. The detection result can accurately locate the block-wise tampered areas.

In practice, it is not enough to only detect whether an image is attacked but it should also be possible to locate the tampered areas and, if possible, recover the original data. Through the location of tampered areas, the attacker's intention can be known and recovery of the tampered parts will follow. The recovered image of the tampered image is shown in Fig. 2.10e where the hat and the mouth of the original image are recovered.

Some state-of-the-art watermarking methods with recovery are listed as follows. Fridrich and Goljan [15] introduced two techniques for self-embedding of an image. The first method is based on transforming 8×8 blocks of an image using DCT quantization coefficients. Thenceforth watermarks are embedded in the LSBs. This approach gives a high quality of reconstructed image but it is very fragile. Another method uses a principal similar to differential encoding to embed a circular shift of the original image with a decreased color depth into the original image. Although the reconstructed image gradually degrades with the increasing amount of noise, it is more robust. An adjacent-block based statistical detection approach to accurately identify the tampered block was proposed by He et al. [16]. It is a fragile watermarking algorithm in tamper detection under collage attack and content-tampering attack. Lee and Lin [17] proposed a dual watermarking scheme for image tampering detection and recovery. Each block in the image contains a watermark of an additional two blocks. This means that there are two copies of the watermark for each nonoverlapping block in the image. Therefore, the scheme maintains two copies of the watermark of an entire image and provides a second chance for block recovery

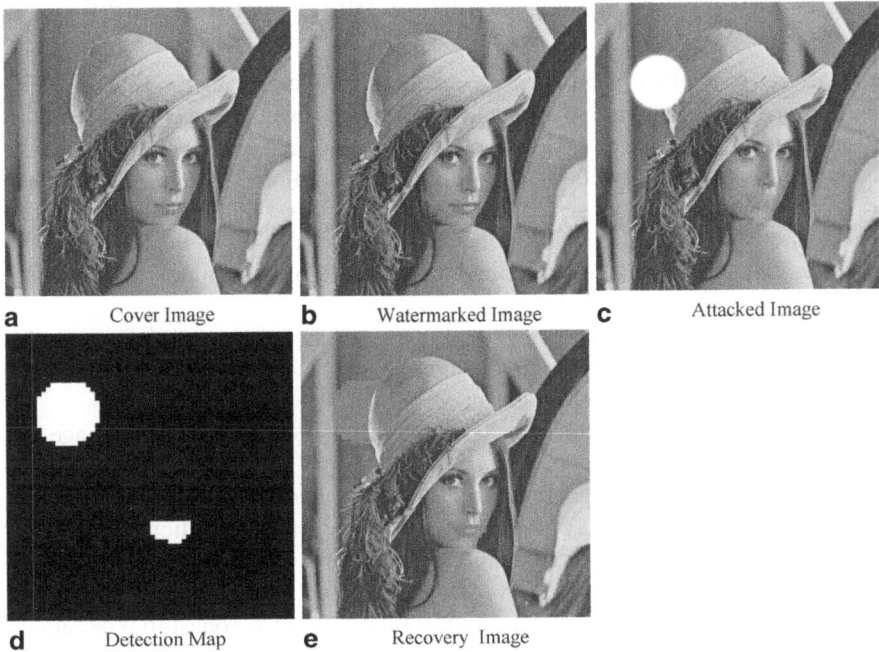

Fig. 2.10 An example of watermarking image authentication with error location

in case one copy is destroyed. This method is robust against large tampered area but as the watermark is embedded in the LSB it is fragile for legitimate operations. Data about the principal content in a region is always embedded into another region for content recovery at the receiver. When both the certain region and the region containing its original information are tampered, the content recovery fails. To deal with this problem, a hierarchical mechanism combined with an exhaustive search method for retrieving the original most significant bits (MSB) was introduced by Zhang and Wang [18].

Nevertheless, this approach is effective only if the tampered area is smaller than 6.6 % of the whole image. Zhang et al. [19] proposed a method, in which the watermarking data produced by exclusive-or operations on the original MSB of a pair of pixels is embedded into the LSB planes. The receiver may estimate the pixel values based on the original or recovered neighboring pixel values. If the tampering percentage is high, recovery will still fail. Zhang et al. [20] proposed a novel self-recovery watermarking scheme. The watermarking data for content recovery is calculated from the original DCT coefficients of the host image. When a part of the watermarked image is tampered, the un-tampered watermarking data can still be extracted using compressive sensing (CS) technique or compositive reconstruction.

2.4 A Watermarking Algorithm for Image Authentication

As stated before, there are two main image authentication methods, i.e., digital signatures and digital watermarking. Both of them have their own pros and cons. Using digital signatures for data authentication is quite mature and is already in wide use but it needs extra bandwidth. Due to the use of a hash function, they are susceptible to failed authentication because of the avalanche effect. Digital watermarking can realize authentication without the aforementioned shortcomings. Since it is difficult to be discarded or tampered, watermarking technology is gaining wide interest in recent researches.

Just like digital signatures, the fragile watermarking approach is proposed to verify the integrity and authenticity of digital content and location of tampered or modified areas using the embedded data.

In practice, an image will always be processed using some common image processing operations such as compression and filtering. The fragile watermarking is not suitable for these situations, so the semi-fragile watermarking technique has been proposed. Lin and Chang [21] proposed a novel image authentication system based on semi-fragile watermarking technique. It is the first system which has the capability of distinguishing malicious attacks from acceptable operations. It is sensitive to detection and location of malicious manipulations but is robust to a certain extent against JPEG lossy compression. Cheddad et al. [22] presented a semi-fragile watermarking method using a halftone version. It uses imperceptible information hiding technique to add another security layer which is resistant to noise and JPEG compression. Chamlawi et al. [23] proposed a wavelet domain based semi-fragile watermarking scheme. It exhibits robustness to JPEG compression. However, the tolerable tampering percentage is not high. Zhu et al. [24] presented a semi-fragile watermarking approach for automatic authentication and restoration of the content of digital images. They formulate the restoration problem as an irregular sampling problem where tampered blocks are detected. Their method is robust to common image processing operations such as lossy transcoding and image filtering but their acceptable tampered areas are small.

In this section, designing watermarking for image authentication is introduced and an example of semi-fragile watermarking is given.

2.4.1 Watermark Design

As mentioned above, a watermark is always generated using the features of the cover image, which will be embedded into the host image. If some parts of the watermarked image are manipulated, they can be detected via the extracted watermark.

In this section, a recent watermarking approach is briefly introduced (see Fig. 2.11).

Suppose that the cover image is divided into 10×10 blocks. First, the features for the cover image are calculated and then the watermark is generated. Here, wm_A,

Fig. 2.11 Watermarking embedding

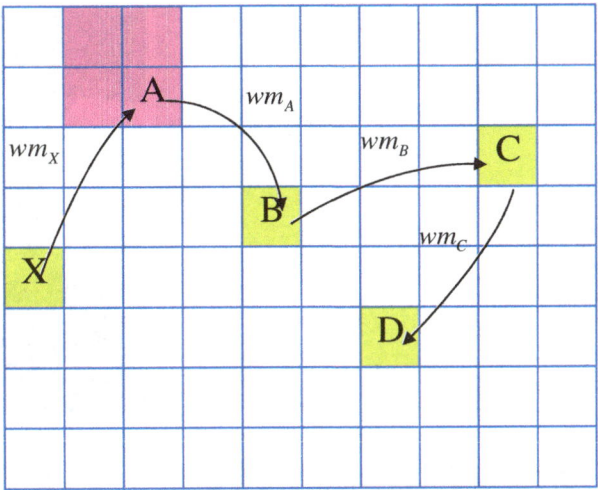

wm_B, wm_C, wm_D, wm_X denotes the corresponding watermarks of block A, block B, block C, block D and block X (yellow blocks). Scrambling/permutation is used to encrypt the original watermark using a secret key. As shown in Fig. 2.12, the extracted watermark wm_A from Block A is not embedded in the Block A itself, rather it is embedded into the Block B. The watermark wm_B is inserted into Block C and so on. Finally, the watermark wm_X will be embedded into Block A. The reasons for using scrambling are: (1) it introduces de-correlation into the watermarking, thus making it more secure; (2) the image feature and its watermark are not destroyed at the same time.

Since the features of an image are correlated in space, the de-correlation in watermarking makes it hard for attackers to find the original watermark and to change it. For example, if the Block A is tampered (red block) its feature may change. If the watermark wm_A was also embedded in Block A, wm_A would unfortunately be destroyed due to the changes. However, if watermark wm_A is inserted into Block B, which is un-tampered, wm_A will survive. It will detect that the Block A is tampered. Accordingly, the Block X is not falsified, but its watermark wm_X which is embedded into Block A is. Comparing the wm_X with the regenerated wm_X' from the un-tampered Block X, Block A can also be authenticated.

Because the tampered areas are always concentrated, scrambling disperses the corresponding watermarks into the whole image and increases the tamper detection capability.

In general, the watermarking scheme discussed in this section can be divided into two parts. The first part is feature extraction and watermark generation. The second part is watermark embedding and later on its extraction for authentication.

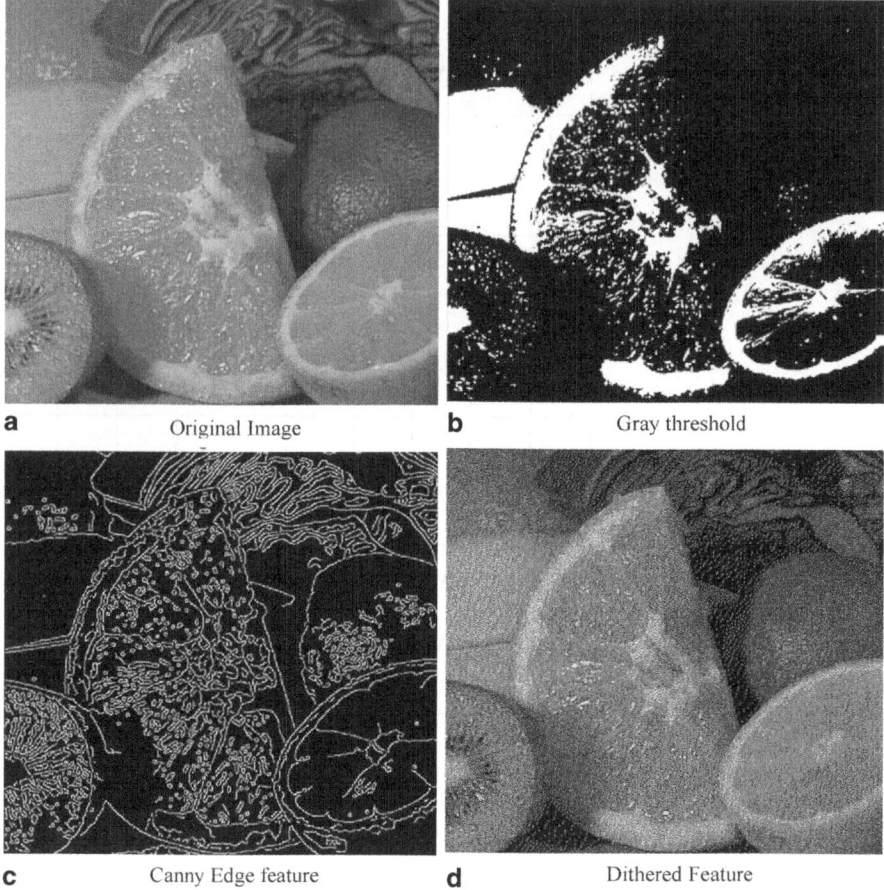

Fig. 2.12 Image in different binary representations. **a** Original image, **b** gray threshold version, **c** edge version, and **d** dithered version

2.4.1.1 Feature Extraction

There are many kinds of interesting image features, such as gray threshold, edge, dithered, statistical, DCT, DWT, etc.

Binary Representations

Image features are essential, as they are embedded into the cover image and therefore the features should be very precise and concise to make them good for embedding. There is a trade-off between perceptual visualization and space demand for embedding. Without taking compression into account, the payload can be consistent with the cover image. If the host image is stored as an 8-bit unsigned integer type, then one pixel will require an 8-bit payload, as shown in Fig. 2.12a. An approximation of the cover image can be achieved by applying the gray threshold technique which results in a binary image demanding only 1-bit per pixel for storage, as shown in Fig. 2.12b. As illustrated in Fig. 2.12c, the edges of objects in an image are very important features, since they reflect the principal contour of the objects in the image. Another approach is dithering as shown in Fig. 2.12d. Dithering is a process by which a digital image with a finite number of gray levels is made to appear as a continuous-tone image.

Statistical Feature

There are many features based on the statistical values of the image data, e.g., histogram of pixel values. In statistics, a histogram is a graphical representation of the distribution of data. It is an estimate of the probability distribution of a continuous variable.

The height of a rectangle is also equal to the frequency density of the interval, i.e., the frequency divided by the width of the interval. A histogram may sometimes be normalized for displaying the relative frequencies. It then shows the proportion of cases that fall into each of several categories, with the total area equaling 1. Figure 2.13 shows histograms of the "house" and "boat" images.

Transform Coefficients

Transform coefficient is another method to obtain the image features. A time domain image shows how a signal changes over time. A frequency domain graph shows how much of the signal lies within each given frequency band over a range of frequencies. The transform coefficients, such as FT, DCT, and DWT, are relatively stable. Usually, the image is divided into nonoverlapping 8×8 blocks. Then, for each block, the DCT coefficients are calculated. Afterwards the DC coefficients are used to generate the watermark, since the DC coefficient in an image block is more stable as compared to the AC ones.

Other Features

There are plenty of other features, for example, pseudo-random sequences and principal component analysis (PCA). Pseudo-random sequence watermark has a random seed that is used to generate the watermarked matrix. This seed must be stored like a secret key and is used in the detection process to for watermark reconstruction. The use of binary watermarks with this algorithm is rather common. Chaotic sequence is a common pseudo random sequence generation method.

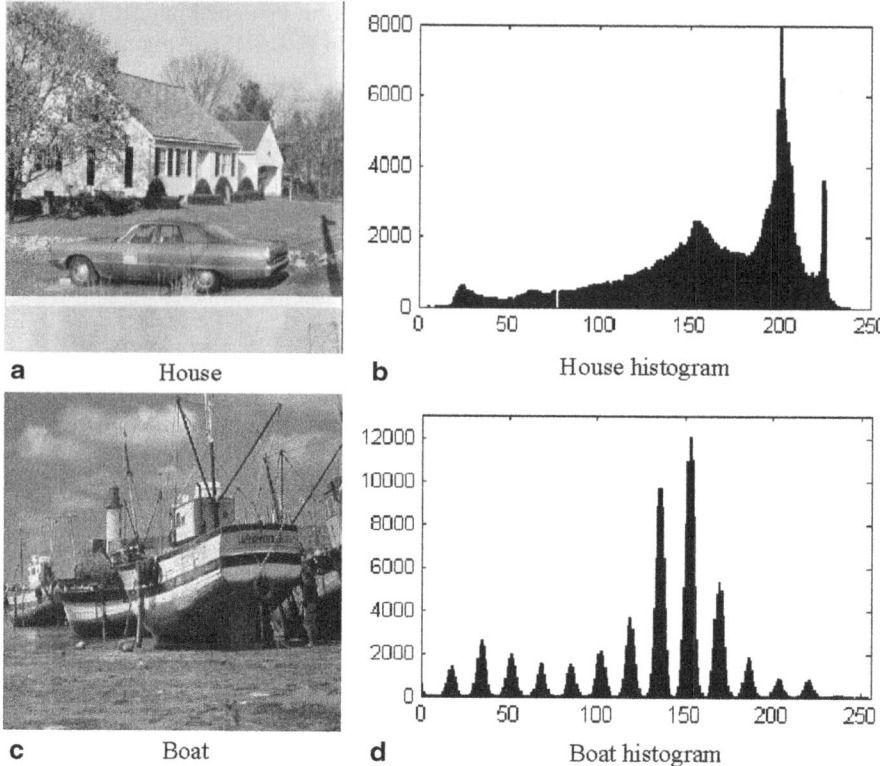

Fig. 2.13 Histograms of different images

An image can be seen as a high dimensional data. If there are $m \times n$ pixels in an image, then each image yields a data point in $R^{m \times n}$. However, human senses do not consider the high-dimensional pixels, and will rather deal with the $m \times n$ pixels to generate a low-dimensional natural structure. The watermark can be generated using the low dimensional data, for which PCA is the most common method. It uses orthogonal transformation to convert a set of observations of possibly correlated variables into a set of values of linearly uncorrelated variables called principal components. The number of principal components is less than or equal to the number of original variables. This transformation is defined in such a way that the first principal component has the largest possible variance (that is, it represents as much variability in the data as possible), and each succeeding component in turn has the highest variance possible under the constraint that it be orthogonal to (i.e., uncorrelated with) the preceding components. PCA can generate a robust feature, thus it is always used in robust watermarking.

2.4.1.2 Watermark Embedding Approach

The watermark embedding approach needs to be considered according to the design goals. In order to design a robust watermarking scheme, a robust method should be used. For example, SS is a good choice for robust embedding. More redundancy needs to be inserted in the watermark for robustness. ECC make a watermark more robust against attacks.

In order to design a fragile watermarking, LSB method can be a good choice. Because of the large capacity and sensitivity, the embedded watermark changes after most manipulations. For the watermark embedding procedure, research has shown that transform domain methods have a robust performance for semi-fragile watermarking methods. Using the transform coefficients, the embedding approach is robust against lossy compression, noise, and many signal processing operations.

2.4.2 An Example Design Case

2.4.2.1 Goals

Some design goals of the proposed algorithm are,

1. Semi-fragile watermarking

This method should be robust to most image processing operations, such as lossy compression, noise, and filtering, but it still needs to be fragile to malicious attacks.

2. Kerckhoffs's principle

The security of an image authentication system should closely rely on the secret key. Suppose that the watermark embedding method is public and the attackers know already that the image has been watermarked. The adversary may extract the original watermark, tamper the image, and then re-embed the original watermark into the tampered image.

2.4.2.2 Hybrid Feature Watermark Generation

Assume that the luminance component of the image is Y and the image has N_1 columns and N_2 rows of pixels (both N_1 and N_2 are multiples of 8). Thus the total number of pixels is $N = N_1 \times N_2$. $Y(x, y)$ denotes the pixel, where $1 \leq x \leq N_1$ and $1 \leq y \leq N_2$. Each pixel $Y(x, y)$ has the range of gray levels as $[0, 255]$. For a watermarking content authentication algorithm, effective feature is very important. If the extracted feature represents the main semantic information, it is capable to be resistant against the legitimate signal operations. The feature will be constant if the semantic information does not change. In this example, the hybrid feature focuses on the main semantic information of the images. The first part is a gray threshold feature

GT, where the threshold is an adaptive median *med*. The gray threshold feature can be calculated as follows:

$$GT = \begin{cases} 1 & if \quad Y(x,y) \geq med \\ 0 & if \quad Y(x,y) < med \end{cases}$$

$$med = median(Y) \tag{2.11}$$

where $median(Y)$ is the function of finding the median of Y.

The other part is the edge detection based feature extraction, where the edge feature EG can be obtained using known edge detectors, such as the Canny edge detector. The edge based feature represents one of the most important and widely used feature of an image.

When the original hybrid feature is generated, the amount of data is too large to be embedded into the cover image. In this example case, only 8 × 8 DCT blocks are considered, so two features GT and EG should be divided into blocks. 2 bits are used to represent GT and EG. The blocking functions are as follows:

$$GT_{(i,j)} = \begin{cases} 1 & if \quad \sum_{k=1}^{8\times 8} GT_{(i,j)}^k \geq \tau_{gt} \\ 0 & if \quad \sum_{k=1}^{8\times 8} GT_{(i,j)}^k < \tau_{gt} \end{cases} \tag{2.12}$$

$$EG_{(i,j)} = \begin{cases} 1 & if \quad \sum_{k=1}^{8\times 8} EG_{(i,j)}^k \geq \tau_{eg} \\ 0 & if \quad \sum_{k=1}^{8\times 8} EG_{(i,j)}^k < \tau_{eg} \end{cases} \tag{2.13}$$

where (i,j) denotes the indices of non overlapping 8 × 8 blocks. $GT_{(i,j)}^k$ and $EG_{(i,j)}^k$ are the k-th GT and EG coefficients of the (i,j) 8 × 8 blocks respectively. $GT_{(i,j)}$ and $EG_{(i,j)}$ denote the output GT and EG feature for each 8 × 8 block. τ_{gt} and τ_{eg} are predefined thresholds. In other words, blocking functions convert the hybrid feature into 2-bits. After blocking, the size of GT and EG becomes N respectively. GT and EG will compose the hybrid feature HF, which is the 2 bit hybrid feature for each 8 × 8 DCT block. The LSB is a gray threshold feature GT, and the MSB is the edge feature EG. In order to increase the security, the torus automorphism is used for permutation using a secret key, written as

$$\begin{bmatrix} i' \\ j' \end{bmatrix} = \begin{bmatrix} 1 & 1 \\ k & k+1 \end{bmatrix} \begin{bmatrix} i \\ j \end{bmatrix} \pmod{S} \tag{2.14}$$

where (i,j) is the original position and (i', j') the new position. S is a positive integer about the range of transformation. k is the random variable generated by a secret key. Finally, the original feature will be transformed and the watermark is generated.

2.4.2.3 Watermark Embedding

The input original cover image Y is segmented into non overlapping 8×8 blocks $Y_b(i, j)$. After DCT transform, the principal diagonal coefficients are chosen to embed the watermark. In other words, the embedding position is between 30 and 41 after the zigzag scan. In this way, the block has good chances of withstanding most of the attacks. The watermark embedding method is described as follows:

If $wm = 1$, then

$$\begin{cases} C_k(u) = C_k(v), C_k(v) = C_k(u) & \text{if } C_k(u) < C_k(v) \\ C_k(u) = C_k(u) + s, C_k(v) = C_k(v) - s & \text{if } C_k(u) = C_k(v) \end{cases} \quad (2.15)$$

If $wm = 0$, then

$$\begin{cases} C_k(u) = C_k(v), C_k(v) = C_k(u) & \text{if } C_k(u) > C_k(v) \\ C_k(u) = C_k(u) - s, C_k(v) = C_k(v) + s & \text{if } C_k(u) = C_k(v) \end{cases} \quad (2.16)$$

C_k denotes the principal diagonal coefficients; u and v represent the embedding position indices of the diagonal DCT coefficients and s is a small constant.

In order to increase the robustness of a watermarking system, an error correcting code is used. According to the embedding position u and v, the proposed method can embed a 6 bit watermark in each 8×8 block. However, only 2 bits represent the hybrid feature, so 4 error correcting bits can be embedded. Here a (6, 2) repetition code is used. For example, if the watermark is {01}, then the encoded watermark is {01| 01| 01}. Thus, at the receiver side, the watermark extraction is more robust using majority logic decoding. After embedding a watermark, the watermarked DCT block is inverse transformed.

The original image and watermarked image are shown in Fig. 2.14a, 2.14c. The hybrid feature is described in Fig. 2.14b. It shows that the proposed algorithm satisfies perceptual invisibility.

2.4.2.4 Watermark Extraction and Authentication Procedure

Suppose that an adversary has tampered some regions in a watermarked image with fake information as shown in Fig. 2.14d. At the receiver, the luminance component is extracted from the watermarked image Y_t. The input is divided into non overlapping 8×8 blocks and the watermark wm' is extracted as follows:

$$\begin{cases} wm' = 1 & \text{if } C_k(u) > C_k(v) \\ wm' = 0 & \text{if } C_k(u) < C_k(v) \\ wm' = -1 & \text{if } C_k(u) = C_k(v) \end{cases} \quad (2.17)$$

$wm' = -1$ means the bit has an error. Because the original watermark has 4 bit redundancies, the wm' can be corrected by the duplicated information. The watermark

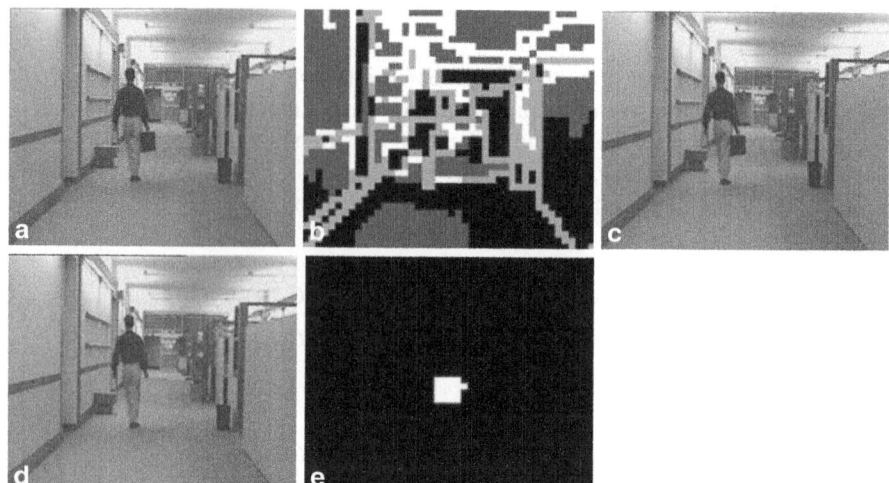

Fig. 2.14 a Original hall image. **b** Hybrid feature. **c** Watermarked hall image (PSNR = 34.57). **d** Tampered watermarked image. **e** Tampered area location map

is inverse permutated and then the extracted hybrid feature HF' is obtained. The watermarked image Y_t generates the hybrid feature $HF\hat{E}^o$ as the watermark embedding procedure. After subtraction, the original tampered area location map TM is created. The original tampered area location map often has some isolated points, so it needs post-processing. After the connected domain computation and some small areas removal, the final tampered area location map is generated in Fig. 2.14e.

References

1. Tirkel AZ, Rankin GA, Van Schyndel RM, Ho WJ Mee NRA. Osborne CF. Electronic watermark. Digital Image Computing, Technology and Applications (DICTA'93); 1993. pp. 666–73.
2. Fung C, Gortan A, Godoy JW. A review study on image digital watermarking. The Tenth International Conference on Networks (ICN 2011); 2011. pp. 24–8.
3. Li C, et al. Multi-block dependency based fragile watermarking scheme for fingerprint images protection. Multimed Tool Appl. 2013;64(3):757–76.
4. Hartung F, Su JK, Girod B. Spread spectrum watermarking: malicious attacks and counterattacks. Image. 1999;1(1):12.
5. Thabit R, Khoo BE. Robust reversible watermarking scheme using Slantlet transform matrix. J Syst Softw. 2014;88:74–86.
6. Arsalan M, Malik SA, Khan A. Intelligent reversible watermarking in integer wavelet domain for medical images. J Syst Softw. 2012;85(4):883–94.
7. Jiménez-Salinas M, Garcia-Ugalde F. Improved spread spectrum image watermarking in contourlet domain. 23rd International Conference Image and Vision Computing New Zealand; 2008. pp. 1–6.

8. Zhang R, Wang H, Wang Y. A novel image authentification based on semi-fragile watermarking. 2012 Fifth International Joint Conference on Computational Sciences and Optimization (CSO); 2012. pp. 631–4.
9. Barton JM. Method and apparatus for embedding authentication information within digital data. US Patent HS5646997, 8 July 1997. 1997.
10. Honsinger CW, et al. Lossless recovery of an original image containing embedded data. US Patent 6,278, 791, 21 Aug 2001. 2001.
11. Tian J. Reversible data embedding using a difference expansion. IEEE Trans Circuits Syst Video Technol. 2003;13(8):890–6.
12. Ni Z, et al. Reversible data hiding. IEEE Trans Circuits Syst Video Technol. 2006;16(3):354–62.
13. Poonkuntran S, Rajesh RS. Chaotic model based semi fragile watermarking using integer transforms for digital fundus image authentication. Multimed Tools Appl. 2012;68(1):79–93.
14. Alavianmehr MA, et al. A semi-fragile lossless data hiding scheme based on multi-level histogram shift in image integer wavelet transform domain. 2012 Sixth International Symposium on Telecommunications (IST); 2012. pp. 976–81.
15. Fridrich J, Goljan M. Images with self-correcting capabilities. 1999 International Conference on Image Processing (ICIP 99); 1999. pp. 792–6.
16. He H, Zhang J, Chen F. Adjacent-block based statistical detection method for self-embedding watermarking techniques. Signal Process. 2009;89(8):1557–66.
17. Lee T, Lin SD. Dual watermark for image tamper detection and recovery. Pattern Recog. 2008;41(11):3497–506.
18. Zhang X, Wang S. Fragile watermarking scheme using a hierarchical mechanism. Signal Process. 2009;89(4):675–9.
19. Zhang X, Wang S, Qian Z, et al. Self-embedding watermark with flexible restoration quality. Multimedia Tools Appl. 2011;54(2):385–95.
20. Zhang X, Qian Z, Ren Y, et al. Watermarking with flexible self-recovery quality based on compressive sensing and compositive reconstruction. IEEE Trans Inf Forensics Secur. 2011;4(6):1223–32.
21. Lin C, Chang S. SARI: self-authentication-and-recovery image watermarking system. Proceedings of the ninth ACM international conference on Multimedia; 2001. pp. 628–9.
22. Cheddad A, Condell J, Curran K, et al. A secure and improved self-embedding algorithm to combat digital document forgery. Signal Process. 2009;89(12):2324–32.
23. Chamlawi R, Khan A, Usman I. Authentication and recovery of images using multiple watermarks. Computer Electr Eng. 2010;36(3):578–84.
24. Zhu X, Ho AT, Marziliano P. A new semi-fragile image watermarking with robust tampering restoration using irregular sampling. Signal Process:Image Commun. 2007;22(5):515–28.

Chapter 3
Perceptual Image Hashing Technique for Image Authentication in WMSNs

Jinse Shin and Christoph Ruland

3.1 Introduction

The rapid development of mobile communication infrastructure and the recent advances in consumer electronic devices such as a smartphone and a tablet PC have allowed individuals to easily produce, distribute, and enjoy multimedia contents in digital form from anywhere at any time. Additionally, due to the availability of low-cost hardware such as a complementary metal-oxide-semiconductor (CMOS) camera and a digital microphone, wireless multimedia sensor networks (WMSNs) which enable to deliver multimedia data in wireless sensor networks (WSNs) have been grown and widely used for a variety of application (e.g., surveillance system, traffic enforcement system, personal health monitoring system) in recent years [2]. However, the same technologies that have brought about profound changes and new opportunities in our lives have also enabled people to easily manipulate and duplicate multimedia data without any trace. In this context, a secure multimedia communication becomes increasingly more important than before since visual data may contain sensitive information which should be always authentic. Otherwise, tampered data or malicious data may lead to a wrong decision and cause serious problems for the public and each individual in many ways.

The simplest way to provide data authentication mechanism is to directly employ a traditional security solution based on cryptography. A data authentication scheme using a cryptographic approach can be largely categorized into two types according to the data integrity criteria: hard authentication and soft authentication [30].

J. Shin (✉) · C. Ruland
Chair for Data Communications Systems, University of Siegen,
57076, Siegen, Germany
e-mail: jinsuh.shin@uni-siegen.de

C. Ruland
e-mail: christoph.ruland@uni-siegen.de

A hard authentication approach does not allow any bit changes on the input data, like a standardized cryptographic scheme such as message authentication code (MAC) or digital signature standard (DSS) [13, 25]. To do this, a hard authentication scheme generates an authentication code from the input data and delivers it along with the data. During the verification process, the authentication code recalculated from the received data are expected to exactly match with the received authentication code as long as the received data or authentication code are not corrupted or manipulated in transit.

On the other hand, a soft authentication approach is more flexible to small changes on the input data in such a way that it can tolerate a certain level of noise and error caused by an error-prone wireless channel. This noise and error resilience property can be achieved by measuring the distance between the recalculated authentication code and the received one, and then comparing the difference with a preset threshold. As a consequence, a soft authentication scheme can conclude that the received data is authentic when the difference is smaller than the threshold. Otherwise, it is declared as non-authentic. In recent years, researchers have proposed several different ideas, such as approximate message authentication code (AMAC), noise tolerant message authentication code (NTMAC), and fuzzy authentication, in order to deal with the challenging problems of data authentication in the presence of noise [3, 7, 30].

Although a soft authentication approach can have the error resilience property, there are still some limitations for secure multimedia communication. The main challenge is caused by the perceptual redundancy of multimedia data which cannot be noticed by human. For instance, a sensor node in WMSNs may apply image compression before data transmission because it is the most effective way to decrease the required energy consumption and network bandwidth by reducing the size of image data to be transferred [2, 6]. However, a general image compression process mainly aims at removing the perceptual redundancies of an image so that it does not affect image content itself while reducing the size of image data and having completely different pixel values compared to the original image. Besides image compression, there are several acceptable image processing operations which keep the same perceptual meaning of the original image, but generate a totally different binary representation. Accordingly, a soft authentication scheme cannot provide enough robustness against such an incidental distortion caused by an acceptable image processing operation (e.g., lossy compression) nor locate a tampered region when image content is maliciously manipulated, because soft authentication is also basically designed to authenticate the binary representation of the input data. Therefore, the existing data authentication schemes based on cryptography are not practically suited for secure multimedia communication even though they have been provided an adequate solution for legacy networks.

To cope with the above mentioned challenges of WMSNs, a digital watermarking based image authentication approach has emerged as an alternative data authentication scheme due to the robustness and the imperceptibility of the digital watermarking technique [11]. Although the digital watermarking technique can deal with those challenges, the extra source coding to embed a watermark into an image imposes significant overheads on a sensor node which has constraints on the limited energy and computational resources. For this reason, this chapter introduces a content based image authentication approach using perceptual image hashing technique which is receiving an increased attention in multimedia security domain. Moreover, this chapter also investigates the feasibility of the use of perceptual image hashing technique as another alternative for image authentication in WMSNs.

The rest of this chapter is organized as follows: Sects. 3.2 and 3.3 respectively cover the high level requirements for image authentication in WMSNs and the previous works regarding data authentication schemes for WMSNs. Section 3.4 explains the basic concept and three desirable properties of perceptual image hashing. In addition, the detailed descriptions of the existing perceptual image hashing algorithms are reviewed in Sect. 3.5, followed by Sect. 3.6 presenting the experimental results for the selected algorithms with respect to robustness, discriminability, and security. Finally, the last section concludes this chapter with a brief summary.

3.2 High Level Requirements for Image Authentication in WMSNs

Image authentication has been studied for more than a decade to handle the problems of image transmission over unsecured wireless networks and to protect the authenticity of an image. However, requirements for image authentication in WMSNs are different in many ways from those of traditional data communication systems. Therefore, high level requirements for image authentication in WMSNs are discussed in this section. Figure 3.1 briefly shows the high level requirements for image authentication in WMSNs.

From a security perspective, the definition of data integrity is the main difference in requirements between WMSNs and traditional data communications. In this regard, the robustness, error resilience, and discriminability requirements should be satisfied in order to provide data integrity which requires a different criterion

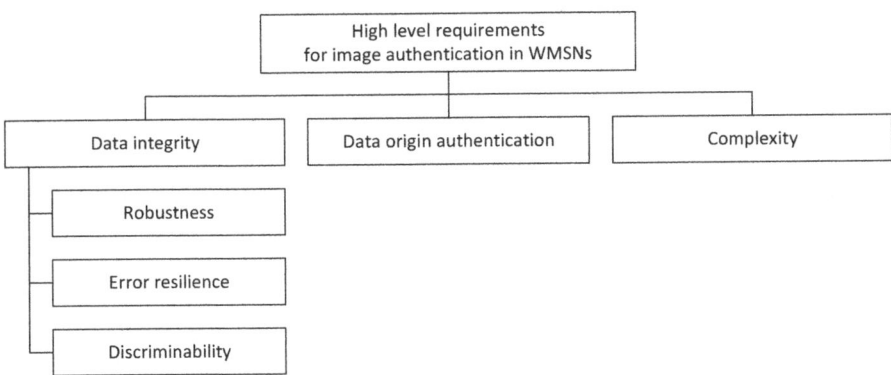

Fig. 3.1 High level requirements for image authentication in WMSNs

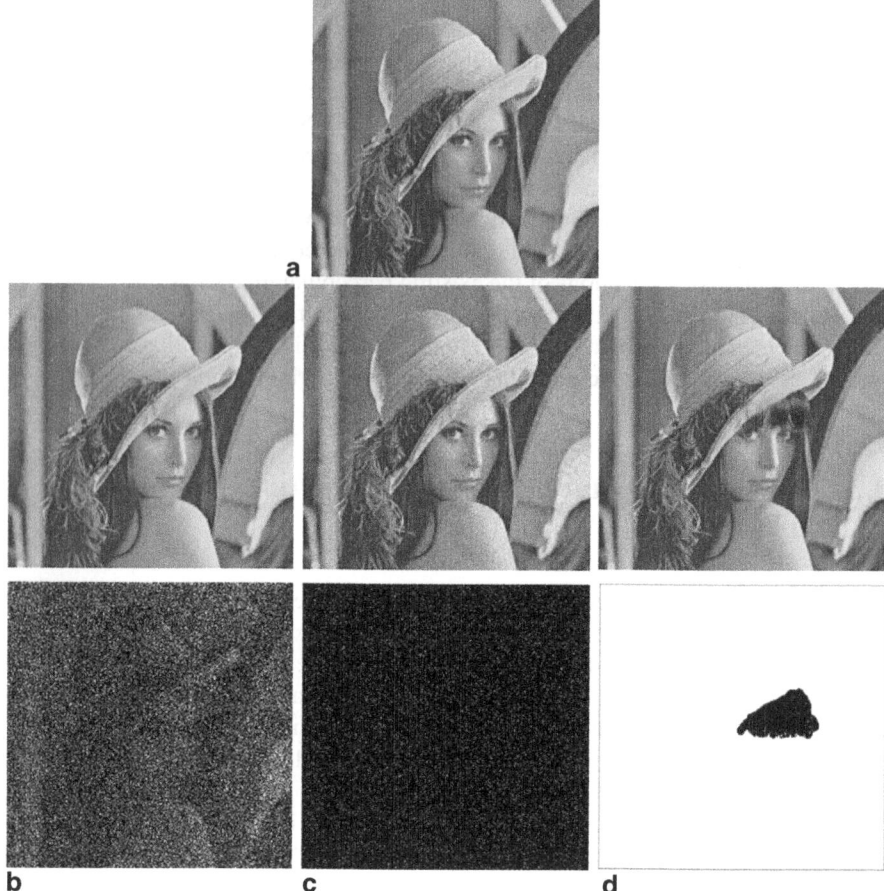

Fig. 3.2 Original Lena image and three distorted versions of Lena image with the binarized absolute difference map between the pixel values of the original and the distorted image. **a** Original image. **b** JPEG compression, $QF = 90$ (PSNR = 40.82 dB). **c** AWGN, $\sigma^2 = 0.005$ (PSNR = 22.99 dB). **d** Tampered image (PSNR = 25.60 dB)

for image authentication in WMSNs. Besides, data origin authentication and low complexity are required as well. Those requirements are described below.

Data Integrity Data integrity is generally defined as the completeness and the consistency of the received data which ensure that data have not been corrupted or modified in its binary representation during a data communication procedure. For image authentication in WMSNs, however, it is desirable to authenticate the perceptual content of an image rather than its binary representation. Thus, no matter what the binary representation of the image has been received, it should be considered as authentic as long as the underlying perceptual content of the image has not been changed. For instance, Fig. 3.2 illustrates three different distorted versions of the

Lena image and the corresponding absolute difference maps between the pixel values of the original image and the distorted ones. As can be shown in Fig. 3.2b and c, both of the distorted images should be declared as authentic since their perceptual content is exactly identical to the original Lena image even though more than 80 and 97 % of the pixel values of the image have been changed respectively. Compared to those content preserving modifications, the tampered image shown in Fig. 3.2d is more similar to the original image in terms of its binary representation since only 3.2 % (8346 pixels out of 512 × 512 total pixels) has been changed in the image.

Robustness In WMSNs, performing multimedia in-network processing (e.g., in-network compression) on the acquired raw data has been considered as one of the solutions to reduce the energy consumption and the amount of data to be transmitted [2]. Furthermore, WMSNs even allow dividing a computationally demanding functionality into several small tasks, and then those tasks can be assigned to a set of sensor nodes in order to efficiently utilize the constrained resources of WMSNs [14]. For a simple example in a video surveillance application, a single captured image can be transferred to the sink node instead of video streaming when any intrusion is detected by the result of object recognition and tracking algorithm. The captured image can be also compressed at any intermediate node. From a security perspective, however, it should be irrelevant what and where multimedia in-network processing has occurred in the end-to-end communication, as long as those operations preserve the perceptual content of an image. Accordingly, image authentication in WMSNs should be robust against acceptable image processing operations preserving the semantic meaning of image content. Thus, the authenticity and the integrity of an image which underwent acceptable image processing operations can be transparent to the end-to-end secure communication, and finally verified as authentic.

Error Resilience Unlike wired networks, packet loss or transmission error may often occur in WMSNs because the wireless link quality is frequently unstable. More importantly, a resource constraint sensor node may cause more packet loss due to the buffer overflow [39]. Although those packet loss and transmission error result in perceptual degradation or damages on the delivered image without any malicious attacks, the received image may still keep the perceptual content of the original image. Therefore, the error resilience property is required for image authentication in WMSNs. Error resilience can be classified as the robustness requirement, but it is separated from the robustness on purpose because of the loss- and error-prone nature of WMSNs.

Discriminability To detect any malicious modifications in an image, it is necessary to provide discriminability which can distinguish content changing modifications from content preserving modifications. The discriminability requirement is often referred as sensitivity in the literature [18]. However, a strong statistical and perceptual redundancy on an image makes it difficult to distinguish malicious manipulations especially when the size of a manipulated area is relatively small. For example, the peak signal-to-noise ratio (PSNR) of the tampered Lena image shown in Fig. 3.2d is even larger than the Lena image with additive white Gaussian noise (AWGN) of

$\sigma^2 = 0.005$. It means that the distortion of the tampered image in this example is smaller than the noisy image that contains perceptually identical image content. Moreover, the discriminability requirement is in conflict with the robustness and error resilience requirements. Thus, the tradeoff between those conflicting requirements should be carefully taken into account in order to design a practical image authentication system in WMSNs depending on the purpose of the application.

Data Origin Authentication One of the main obstacles in WMSNs is that sensor nodes are generally deployed in an unattended environment and managed remotely. In other words, all the sensor nodes are exposed to a physical attack so that an attacker can take over a sensor node or even deploy his/her own sensor nodes in WMSNs. Accordingly, it is highly possible for an attacker to fool the network into accepting manipulated messages or malicious messages injected from his/her own sensor nodes [32, 33]. For this reason, image authentication in WMSNs should provide a mechanism which a recipient can make sure that the received image originates from the claimed source node [27].

Low Complexity Processing image data requires typically more resources in terms of CPU usage, memory, and network bandwidth than processing scalar data. Additionally, the image authentication functionality also imposes a computational overhead on a resource constrained sensor node [2, 37]. Thus, the computational complexity of image authentication should be considered to realize low energy consumption and a low cost device in WMSNs.

3.3 Previous Works on Data Authentication in WMSNs

As a traditional way to ensure data integrity and data origin authentication, WSNs employ a cryptographic approach. In particular, wireless scalar sensor networks (WSSNs) commonly utilize a symmetric cryptographic algorithm rather than an asymmetric cryptographic algorithm since an asymmetric cryptographic algorithm is known to require more processing power and memory resources than what typical sensor nodes can offer. Accordingly, data authentication schemes using an asymmetric cryptographic algorithm are considered as not suitable for WSNs due to the resource constraints of sensor nodes.

Compared to typical sensor nodes in WSSNs, multimedia sensor nodes are generally equipped with sufficient computational resources in order to properly process multimedia data. Thus, the use of an asymmetric cryptographic algorithm can be considered as a feasible solution for WMSNs [8]. However, both symmetric and asymmetric algorithms still leave some issues because of the distinct requirements already explored in Sect. 3.2. The primary challenge is the robustness and error resilience requirements which cannot be supported by cryptographic algorithms. Additionally, since image data are generally very large in size, applying a cryptographic algorithm on a large amount of data incurs a significant overhead even though sensor nodes for WMSNs become powerful enough to satisfy the high resource requirements [8, 17]. For those reasons, a digital watermarking based approach has been investigated in recent years as an alternative solution to provide multimedia data authentication in WMSNs [12, 29].

In [38], Zhang et al. proposed a compression supportive authentication scheme based on the digital watermarking technique to provide the end-to-end authentication as well as in-network compression. To do this, each sensor node embeds a part of the whole watermark into its sensory data by performing a simple mathematical operation, whereas a sink node carries out the watermark verification process which requires a heavy computational load. In this manner, both end-to-end authentication and in-network compression can be achieved by validating the presence of the watermark which may tolerate a certain level of distortion thanks to the robustness property of the digital watermarking technique. Meanwhile, Wang et al. [34, 35] proposed an adaptive energy-aware watermarking scheme to provide a secure image transmission mechanism with energy efficiency in WMSNs. In this scheme, watermarking positions are dynamically chosen by two adaptive thresholds in order to embed a watermark according to the network conditions measured by the packet loss ratio. To increase robustness against packet loss in an error-prone wireless environment, this scheme embeds watermark redundancies into an image and improves the quality of the watermarked image transmission by allocating extra network resources. Hence, it can achieve the energy efficiency and security at the same time. The use of the digital watermarking technique to design a secure communication framework for WMSNs has been also investigated in [11, 21]. Additionally, Harjito et al. [12] presented the evaluation of the existing state-of-the-art watermarking algorithms developed for WMSNs.

A digital watermarking based approach has been proposed as a way to satisfy the robustness requirement for image authentication in WMSNs. However, the main drawback of the use of digital watermarking technique is the fact that it requires extra source coding which incurs a significant overhead on sensor nodes despite of their sufficient computational resources. The low watermark information hiding capacity can be a serious problem as well. In this regard, the next sections will take a look at another approach which utilizes the perceptual image hashing technique to authenticate an image based on the perceptual image content.

3.4 Perceptual Image Hashing

In this section, the basic concept and the desirable properties of perceptual image hashing which can meet the requirements discussed in Sect. 3.2 are discussed.

3.4.1 Basic Concept

Similar to a cryptographic hash function, a perceptual image hash function is designed to take an image as an input and produce a fixed-length output, which is called a hash value or message digest. Unlike a cryptographic hash function, however, the basic idea behind perceptual image hashing is to compute a hash value from the perceptual image content rather than its binary representation. To do this, a perceptual image hash function requires extracting the invariant perceptual features of a given image, and then constructing a hash value from the extracted features. Another main difference between a classical cryptographic hash function and a perceptual image hash function is the verification criterion. A perceptual image hash function measures the distance between two hash values and compares the difference with a preset threshold in order to provide the robustness property, whereas a cryptographic hash function requires an exact matching to ensure the input data has not been altered.

By generating a hash value from perceptual features and employing the threshold based verification, perceptual image hashing can achieve a certain level of robustness against acceptable image processing operations and transmission error. Accordingly, the perceptual image hash technique is receiving an increased attention in the multimedia domain for image identification as well as image authentication.

3.4.2 Desirable Properties

There are three desirable properties required for a perceptual image hash function to possess for the purpose of image authentication [23]. Before discussing the properties of a perceptual image hash function, the following notations should be defined for a clear understanding: I denotes a particular image, and the original image of the given image I is denoted by I_{origin}. Similarly, I_{ident} denotes a perceptually identical image which looks same as the original image while I_{diff} represents a perceptually distinct image. K denotes the key space of all possible keys k_i. Let $H(I, k_i)$ represents a perceptual image hash function with a secret key k_i, which produces a l-bit hash value h for the given image I. Additionally, $D(h_1, h_2)$ is a distance function to measure the similarity or distance metric between two hash values, h_1 and h_2. The measured distance metric requires to be compared with a threshold θ to consider how similar or dissimilar they are.

3.4.2.1 Perceptual Robustness

To satisfy both the robustness and error resilience requirements mentioned in Sect. 3.2, a perceptual image hash function requires producing approximately the same hash value when two perceptually identical images are taken as an input with the same secret key. The following Eq. 3.1 represents the criterion of the perceptual robustness property.

$$Pr\left[D\left(H\left(I_{origin}, K_i\right), H\left(I_{ident}, K_i\right)\right) \leq \theta\right] \approx 1 \quad (3.1)$$

As shown in the above equation, the perceptual robustness property requires that a pair of perceptually identical images should have a distance smaller than a threshold

θ with a high probability. In this manner, a perceptual image hash function can provide tolerance to acceptable image processing operations or transmission errors as long as two compared images keep the same perceptual content.

3.4.2.2 Fragility to Visual Distinct Image

Similar to a cryptographic hash function, a perceptual image hash function also requires generating a different hash value from the different input data. However, a perceptual change on the input data is only considered as a difference for a perceptual image hash function. The following Eq. 3.2 represents the criterion of the fragility property, which means a pair of perceptually distinct images should have a distance larger than a threshold θ with a high probability.

$$Pr\left[D\left(H\left(I_{origin}, K_i\right), H\left(I_{diff}, K_i\right)\right) \geq \theta\right] \approx 1 \quad (3.2)$$

Thus, a perceptual image hash function can distinguish any perceptually distinct image from perceptually identical images so that it can satisfy the discriminability requirement mentioned in Sect. 3.2.

3.4.2.3 Unpredictability of the Hash

The unpredictability of the hash refers to the similar definition as the confusion property in cryptography, which makes it difficult for an attacker to estimate the relationship between a secret key and a hash value even if the algorithm used is known. To provide the unpredictability of the hash, a perceptual image hash function requires that the collision probability resulting in the same output hash value should be approximately $\frac{1}{2^l}$, which means a very low collision probability. The following Eq. 3.3 represents the criterion for the unpredictability of the hash.

$$Pr\left[H\left(I_{origin}, K_1\right) = H\left(I_{origin}, K_2\right) = \alpha\right] \approx \frac{1}{2^l}, \quad \forall \alpha \in \{0, 1\}^l \quad (3.3)$$

Thus, perceptual image hashing can use the randomness obtained from the unpredictability of the hash as its security aspect.

3.5 Content Based Image Authentication Using Perceptual Image Hashing Technique

Content based image authentication is one of the most promising image authentication techniques to authenticate an image in a semantic level. Since a perceptual image hash function should be designed to possess three desirable properties mentioned in Sect. 3.4.2 by obtaining a hash value from the invariant features of an image, it can

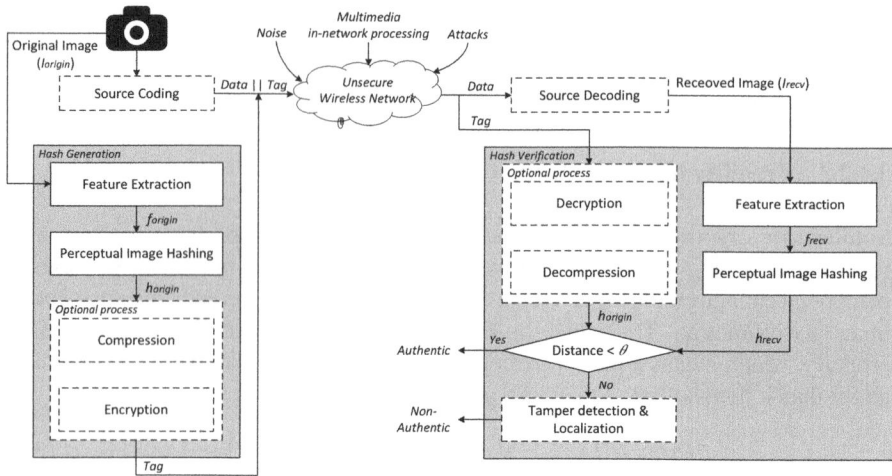

Fig. 3.3 General procedure of content-based image authentication using perceptual image hashing technique

be utilized as a key component of content based image authentication. Figure 3.3 illustrates briefly how a perceptual image hash function works as a component of content based image authentication.

As for the hash generation stage, the invariant features of an original image (f_{origin}) are firstly extracted from I_{orign}. Afterwards, the hash value of the original image (h_{origin}) is calculated based on f_{origin} by applying a perceptual image hash function. Additionally, this stage may include some optional processes to generate a final Tag depending on the algorithm used. For instance, a compression process can be applied to reduce the amount of feature information or the size of the hash value while encryption or random quantization can be processed to protect the hash value or to give a high level of randomness with respect to security. As a result, the final Tag constructed from h_{origin} needs to be transferred along with the image data to a recipient in order to facilitate the verification procedure for the image authenticity and integrity.

Similar to the hash generation stage at a sender side, a recipient requires computing a hash value (h_{recv}) from the received image (I_{recv}). To check the authenticity and integrity of I_{recv}, h_{recv} is compared with the original hash value h_{origin} obtained by applying decryption or decompression on Tag received along with I_{recv}. By measuring the distance between those two hash values and comparing the difference with a preset threshold θ, it may conclude if I_{recv} is authentic or not. Optionally, a tamper detection and localization process may be performed to identify the manipulated area when I_{recv} is considered as non-authentic.

Monga et al. classified the perceptual image hashing technique into four different categories in [23] according to the feature extraction method: Image statistics, relations, coarse image representations, and low-level image representations based feature extraction. Several representative algorithms have been selected from different categories and reviewed in this section. Additionally, some of the selected algorithms are evaluated in Sect. 3.6. For more information on state-of-the-art perceptual image hashing algorithms, the reader may also be referred to [9, 10].

3.5.1 Image Statistics Based Approach

An image statistics-based approach extracts feature vectors from the statistics such as mean, variance, and higher moments of intensity values of an image. In this subsection, the perceptual image hashing algorithm using the statistics of discrete wavelet transform (DWT) coefficients and approximate image message authentication code (IMAC) using the most significant bit (MSB) of 8×8 block average are looked at in detail.

Image Hashing Algorithm Using the Statistics of DWT Coefficients [31] Venkatesan et al. proposed a robust image hashing algorithm that utilized image statistics—either averages or variances depending on the sub-band—from small rectangles created by a random tiling of each sub-band in the wavelet decomposition of an image. The main steps of this algorithm to generate a final image hash value are presented in the following. Additionally, Fig. 3.4 illustrates how to compute the image statistics of DWT coefficients in this algorithm.

1. Compute the DWT of an input image.
2. Divide randomly each sub-band into small rectangles.
3. Calculate the statistics of each rectangle as the invariant features of a given image.
   a) Calculate the average of coefficients a_i ($1 \leq i \leq M$, where M is the number of rectangles) in each rectangle from the approximation sub-band, and obtain F_{LL} by concatenating the results.
   b) Calculate the variance of coefficients, v_i ($1 \leq i \leq M$, where M is the number of rectangles) in each rectangle from the other sub-bands, and obtain F_{LH}, F_{HL}, F_{HH} respectively.
4. Concatenate all the image statistics vectors, and quantize them randomly using a randomized quantizer.

The proposed algorithm was able to achieve better robustness against several content preserving modifications by using the statistics of DWT coefficients rather than using the image intensity directly. However, it is still not robust enough against some modifications that change the contrast and brightness of an image. Furthermore, this algorithm possesses a security weakness that images can be easily modified without altering their image statistics.

IMAC Using the MSB of 8×8 Block Average [36] Unlike other perceptual image hashing algorithms, IMAC employed a cryptographic primitive called AMAC [7]. Besides, IMAC utilized the MSB of 8×8 block average as the invariant features to overcome the limitation of AMAC for image authentication. The main steps of the IMAC generation procedure are shown in the following.

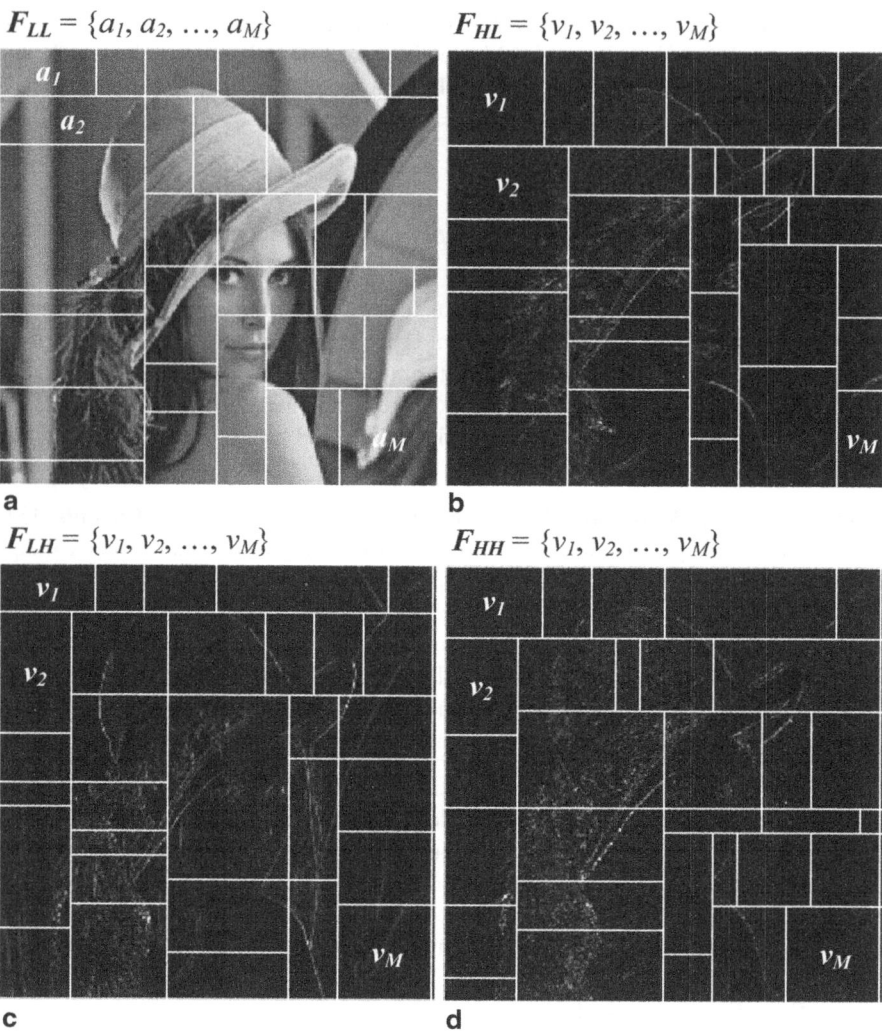

Fig. 3.4 Image statistics vectors from the wavelet decomposition of Lena image. **a** Approximation sub-band (LL). **b** Horizontal detail sub-band (HL). **c** Vertical detail sub-band (LH). **d** Diagonal detail sub-band (HH)

1. Compute 8×8 block average of an input image.
2. Extract the MSB from each 8×8 block average, and then construct the binary map from the result.
3. Calculate a single IMAC bit by applying AMAC on each row and column of the binary map, and concatenate all the IMAC bits.
4. Permute the result, and then perform XOR operation with pseudorandom bits to prevent an attacker from being able to identify any bit of the final IMAC with a specific row or column of binary map.

Since IMAC utilizes the MSB of 8 × 8 block average as the invariant features, it is very sensitive to small perturbations around the average image intensity of 127.5, where the MSB of the block average may change by even a small intensity change on the image. To solve this problem, the authors introduced a guarding zone and image histogram transformation, which improves the error tolerance of IMAC. A guarding zone in conjunction with image histogram transformation can prevent average intensity values from being around 127.5 by splitting the interval of intensity values into two separate regions, $[0, 127 - T_d]$ and $[127 + T_d, 255]$, where T_d is the acceptable maximum absolute difference between the original image and the modified image. In this manner, IMAC can increase its robustness performance. However, this error tolerance enhancement scheme requires modifying the original image before transmission. Moreover, IMAC is still vulnerable to the intensity changes even though the error tolerance enhancement is applied.

3.5.2 Relation Based Approach

A relation-based approach utilizes the invariant relationship between a pair of transform coefficients. In this subsection, the perceptual image hashing algorithm using invariant relationship between discrete cosine transform (DCT) coefficients proposed by Lin and Chang is described in detail.

Image Hashing Algorithm Using Invariant Relationship Between DCT Coefficients [19] Lin and Chang proposed a robust image authentication method based on the invariance of the relationship between DCT coefficients at the same position in separate blocks of an image. This invariant property is derived from the fact that all DCT coefficients at the same position in DCT blocks are divided by the same quantization table for the JPEG lossy compression process. Thus, the proposed algorithm is able to differentiate JPEG compression from malicious manipulations. The following presents how this algorithm generates an image hash value.

1. Compute DCT coefficients for all 8 × 8 blocks of an input image.
2. Form DCT blocks into pairs in order to analyze the relationship of each pair.
3. Analyze the relationship between DCT coefficients in each pair of DCT blocks.
   a) Compute the difference between DCT coefficients at the position v in a pair of DCT blocks p and q.
   b) Generate one bit result according to the following condition.
   $$Z(v) = \begin{cases} 1, & if \ F_p(v) - F_q(v) \geq k \\ 0, & if \ F_p(v) - F_q(v) < k \end{cases},$$
   where k presents a threshold, v is the index of DCT coefficients (in the zigzag order) in a DCT block. F_p and F_q are the DCT coefficient vectors of DCT blocks p and q respectively.
   c) Iterate steps (3a) and (3b) over all the pre-selected positions of DCT coefficients and all the pairs of DCT blocks.

Although this method showed excellent robustness against JPEG compression, some limitations still exist. For instance, most of the AC coefficients in DCT blocks will be zero after JPEG lossy compression, especially in the area where an image has a smooth texture. This result makes the proposed method difficult to extract the valid discriminant features so that it will degrade the performance and lead to a wrong decision. More importantly, this algorithm is vulnerable to defeat-the-scheme attack (DSA) which slightly changes the values of the low and middle frequency DCT coefficients of DCT blocks in the zigzag order while resulting in a significant change in the visual appearance of the image [1]. Other variants of this algorithm published in the literature have not completely solved the above mentioned problems [1, 5].

3.5.3 *Coarse Image Representation Based Approach*

As for the approach based on preserving coarse image representations, feature vectors are extracted from the invariance in the transform domain or coarse image representations. This subsection reviews the perceptual image hashing algorithm using an iterative geometric technique proposed by Mihcak and Venkatesan. Another interesting image hashing algorithm using singular value decomposition (SVD) is also described in this subsection.

Image Hashing Algorithm Using an Iterative Geometric Method [22] Mihcak and Venkatesan proposed to apply a simple iterative filtering operation on the binary map of the DWT approximation sub-band in order to obtain a certain level of robustness. The proposed iterative filtering operation enables to minimize the presence of geometrically weak components and enhance the geometrically strong components by means of region growing. The following shows how this algorithm generates an image hash value.

1. Compute the DWT of an input image.
2. Apply a thresholding operation on the DWT approximation sub-band to produce a binary image M. Note that a threshold value for the thresholding operation requires to be chosen to meet $W(M) \approx q$, where $W(\cdot)$ is the normalized Hamming weight of any binary input, and q is a predefined algorithm parameter indicating the ratio of the usual Hamming weight and the size of the binary input.
3. Apply the geometric region growing algorithm. Let $M_1 = M$.
    a) Apply an order-statistics filtering on M, and obtain M_2
    b) Apply a low-pass linear shift invariant filtering on M_3 via filter f and obtain M_4, where $M_3(i, j) = AM_2(i, j)$ and A is a constant parameter.
    c) Apply a thresholding operation on M_4, and obtain M_5 which meets the condition $W(M_5) \approx q$.
4. Check the following conditions to construct a final image hash value.
    a) Iterate the above procedures (1 – 3) after setting $M = M_5$, until the maximum iteration count is reached or the distance between M_5 and M_1 is less than the threshold ε.
    b) Otherwise, set M_5 as a final image hash value.

This algorithm is designed to use no secret key since it aims to achieve robustness against small modifications. In order to provide a security aspect, the authors proposed to individually apply the above algorithm on randomly partitioned sub-blocks of an input image. By concatenating all the results from each sub-block, a secure image hash value can be constructed. However, the iteration of this algorithm requires a high computational complexity. Likewise, other image hashing algorithms using coarse image representations, the proposed algorithm cannot easily distinguish content changing manipulations from acceptable modifications.

Image Hashing Algorithm Using SVD [16] Kozat et al. proposed an image hashing algorithm, which applies the SVD computation to pseudo-randomly chosen semi-global regions of an image, and then selects the strongest singular vectors to extract robust features. By utilizing the dimensionality reduction technique using SVD, the proposed method was able to achieve the excellent robustness performance against most of content preserving modifications. The step-by-step procedure to construct a final image hash value is outlined in the following and Fig. 3.5.

1. Divide an input image randomly into p possibly overlapping rectangles of the size $m \times m$.
2. Construct the secondary image.
   a) Compute the SVD of each sub-block, and generate $m \times 2$ feature vectors by taking the strongest singular vectors from the results.
   b) Concatenating the results, and construct the secondary image of the size $m \times 2p$ by using a pseudo-random combination of the results.
3. Divide the secondary image randomly into r possibly overlapping rectangles of the size $d \times d$.
4. Compute the SVD of each sub-block of the secondary image again, and generate $d \times 2$ feature vectors by taking the strongest singular vectors from the results.
5. Construct the image hash value by concatenating the results.

By selecting the strongest singular vectors in the SVD of the image, the proposed method was able to increase robustness. However, it may also become a problem since it can lead to misclassification with a high probability for image authentication. More importantly, iterating the SVD computation over each sub-block of both the input image and the secondary image requires a high computational cost. As a variant of this algorithm, the authors also proposed to apply DCT or DWT instead of SVD for generating the secondary image.

3.5.4 Low-level Image Representation Based Approach

A low-level image representation based approach uses low-level image features such as edges or feature points. In general, those low-level image features are highly distinctive in such a way that they have been widely used for the object or scene recognition. Perceptual image hashing algorithms using image edges and feature points are reviewed respectively in the following subsection.

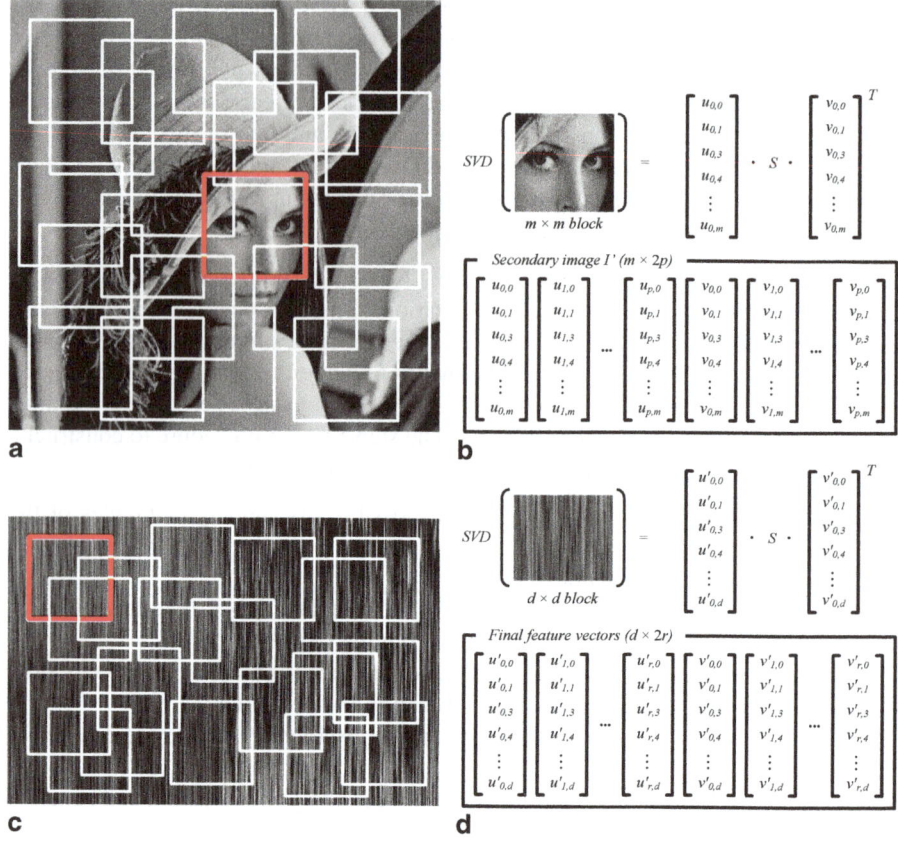

Fig. 3.5 Image hashing algorithm using SVD. **a** Randomly partitioned Lena image. **b** Construction of the secondary image from the strongest singular vectors after SVD computation on each sub-block. **c** Randomly partitioned secondary image of Lena image. **d** Construction of the image hash value

Image Hashing Algorithm Using Image Edges [28] Queluz proposed to utilize image edges for evaluating the integrity of image content as well as detecting malicious manipulations. In the following, the image hash generation procedure is explained.

1. Obtain the binary image of an input image.
    a) Compute the gradient of an input image using the Sobel or Canny operator.
    b) Binarize the results by applying a thresholding operation.
2. Apply a sub-sampling operation to reduce the spatial resolution and simplify image edges.
3. Apply a lossless compression algorithm, and obtain a final image hash.

The main problem of the use of image edges is the fact that edges can be easily distorted by some content preserving modifications. In this regard, the author pointed out that applying JPEG compression with a high compression ratio may cause the smoothing of edges and the mosquito noise, which may create a fake contour. Therefore, the proposed algorithm has some limitations on the robustness performance. Additionally, color manipulations cannot be detected since color manipulations do not change any image edges while changing the image content.

Image Hashing Algorithm Using Feature Points [23, 24] Monga et al. proposed to extract significant image features by using the end-stopped wavelet based feature detection algorithm, which responds strongly to corners and high curvature points in a given image. Afterwards, an iterative procedure is used to lock onto a set of image feature points with a good invariance property to perceptually insignificant perturbations. The step-by-step procedure for this algorithm is described next.

1. Obtain a set of feature points using the end-stopped wavelet based feature point detector, and collect the magnitudes of the wavelet coefficients at the selected feature points to form a feature vector f.
2. Binarize f by applying a probabilistic quantization scheme, and obtain a binary string f_b^1.
3. Apply a geometric region growing algorithm to minimize the presence of geometrically weak components and enhance geometrically strong components.
   a) Perform an order-statistics filtering on the input image, and obtain I_{os}.
   b) Perform a low-pass linear shift invariant filtering on I_{os}, and obtain I_{lp}.
4. Repeat steps (1) and (2) with I_{lp}, and obtain a binary string f_b^2.
5. Check the following conditions to construct an image hash value.
   a) Iterate the above procedures (1 – 4) after setting $I = I_{lp}$, until the maximum iteration count is reached or the distance between f_b^1 and f_b^2 is less than a threshold ρ.
   b) Otherwise, set f_b^2 as an image hash value.

By employing the iterative feature point detection algorithm based on preserving significant image geometry, the proposed method achieved robustness while having the discriminability property to content changing manipulations. However, this algorithm may not be able to capture some small details on the image since it only utilizes the limited number of feature points retaining the strongest coefficients to construct the image hash. Therefore, small malicious manipulations cannot be detected well. Moreover, its hash generation scheme mainly relies on the invariant positions of feature points. Thus, it is highly possible for an attacker to add or remove a set of feature points for malicious manipulations while maintaining the same strongest feature points of an image, so that it may lead to a false positive authentication.

3.6 Experiment Results

As the representative algorithms of each category, five perceptual image hashing algorithms are selected to evaluate their performance in this section. The followings are the algorithms used for experiments: [16, 19, 24, 31, 36].

Based on three desirable properties discussed in Sect. 3.4.2, robustness, discriminability, and security are considered as the performance criteria in order to evaluate the selected algorithms. Robustness and discriminability are assessed by measuring the distances between the hash values of all the original test images and their distorted versions of images, followed by applying the receiver operating characteristic (ROC) curve analysis. The area under the ROC curve (AUC) demonstrates the accuracy of the selected algorithms concerning robustness and discriminability. An AUC value of ≥ 0.9 is generally considered as "excellent", $0.8 \sim 0.9$ as "good", $0.7 \sim 0.8$ as "fair", and < 0.7 as "poor". As for the security, it is evaluated by measuring the randomness property of a perceptual image hashing algorithm since most perceptual image hashing algorithms use the randomness as their security aspect. Accordingly, the irrelevance between the final hash value and secret key is investigated in this experiment.

The performance of the selected algorithms are evaluated by experiments with 50 original images (512×512 grayscale with 8-bit per pixel) including some classical benchmarking images and image sequences of toy vehicle from USC-SIPI image database[1] and several distorted versions of the given original images. In experiments, five different types of content preserving modifications are considered as acceptable modifications to assess the robustness of a perceptual image hashing algorithm: JPEG compression (Quality factor between 5 and 90%), scaling (Scaling factor between 10 and 200%), Gaussian blurring (Standard deviation between 1 and 5 with the filter size 9×9), AWGN (Variance between 0.02 and 0.10), and gamma correction (Gamma value between 0.2 and 2.0). Additionally, one modification which combines parts of another image with the given original test image is regarded as a content changing manipulation in order to evaluate discriminability. The size of manipulated region is varied in the range of 16×16 and 128×128. The types of modifications and the corresponding detail parameters are summarized in Table 3.1.

3.6.1 Robustness

The robustness of the selected algorithms is presented by investigating the impact of five different content preserving modifications. The following experiments measure AUC values for each type of modifications while the corresponding parameters vary as described in Table 3.1.

Figure 3.6 shows the impact of JPEG compression with the quality factor ranging from 5 and 90%. As can be observed, all the selected algorithms are tolerate enough to JPEG compression with the quality factor down to 5% since all the AUC values for the entire range of quality factor parameters are higher than 0.972. In particular,

---

[1] Available from the website of USC-SIPI (University of Southern California—Signal and Image Processing Institute image database), "http://sipi.usc.edu/database/".

3 Perceptual Image Hashing Technique for Image Authentication in WMSNs

Table 3.1 Types of modification and corresponding parameters

Types of modification		Parameters
Content preserving modification	JPEG compression	Quality factor: 5 %, 10 %, 15 %, 25 %, 50 %, 60 %, 70 %, 80 %, 90 %
	Scaling	Scale factor: 10 %, 20 %, 50 %, 150 %, 200 %
	Gaussian blurring	Filter kernel size: 9×9 Standard deviation (σ): 1, 2, 3, 4, 5
	Additive Gaussian noise	Variance (σ^2): 0.02, 0.04, 0.06, 0.08, 0.10
	Gamma correction	Gamma (γ): 0.2, 0.4, 0.6, 0.8, 1.0, 1.2, 1.4, 1.6, 1.8, 2.0
Content changing modification	Manipulation	Size of manipulated area: 16 × 16, 24 × 24, 32 × 32, 40 × 40, 48 × 48, 56 × 56, 64 × 64, 80 × 80, 96 × 96, 112 × 112, 128 × 128

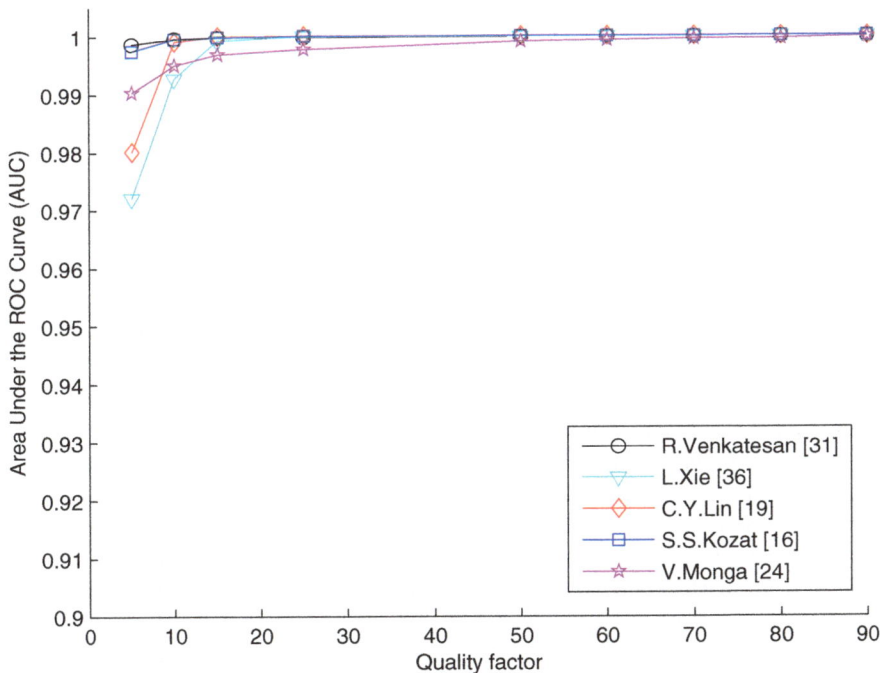

Fig. 3.6 Impact of JPEG compression

Venkatesan [31] and Kozat [16] even reach the AUC value of over 0.997 at the quality factor of 5 %.

A similar impact is observed for scaling and Gaussian blurring in Figs. 3.7 and 3.8 respectively. The AUC values measured from a scaling operation with the scale factor ranging between 10 and 200 % remain over 0.95 even though decreasing the size of

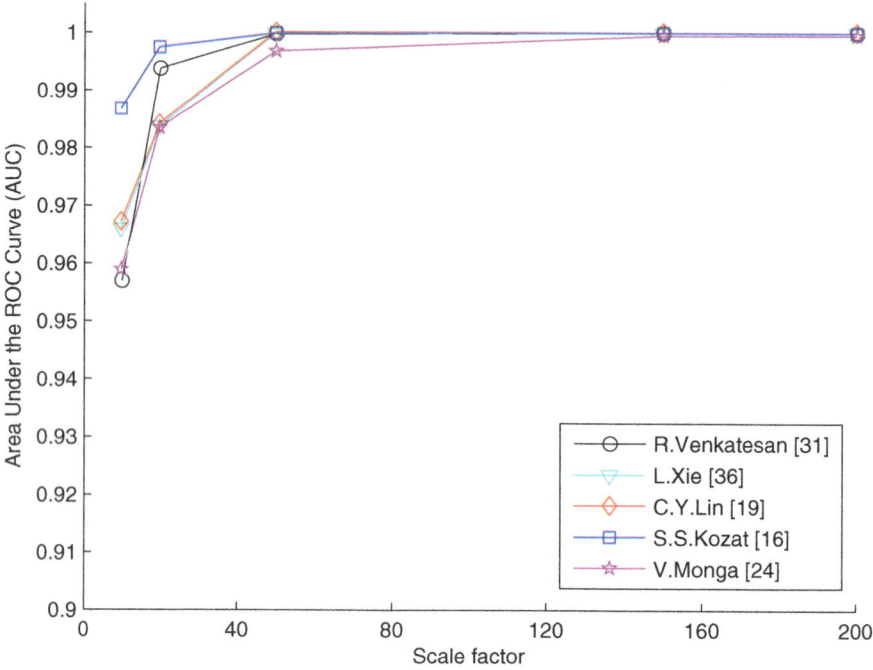

Fig. 3.7 Impact of scaling

the scale factor slightly drops the AUC value for all the algorithms. As for a Gaussian blurring operation, AUC values also decrease as the standard deviation increases, and then finally drop to 0.983, 0.999, 0.991, 0.997, and 0.979 at the standard deviation of 5. Therefore, it can be concluded that all the selected algorithms are robust enough against scaling and Gaussian blurring as well.

Unlike the previous experiments, AWGN and gamma correction affect the robustness of the selected algorithms. As presented in Fig. 3.9, AWGN with the noise variance in the range between 0.02 and 0.1 has a severe impact on the robustness of Venkatesan [31], Lin [19], and Monaga [24]. In the case of Venkatesan [31], it achieves a fair performance by maintaining the AUC value between 0.7 and 0.8 for the entire range of the noise variance. Lin [19], and Monaga [24] begin with 0.961 and 0.934 respectively, and then gradually decrease by up to 0.487 and 0.250. On the other hand, Xie [36] and Kozat [16] stay close to 1 for all the cases which indicates excellent robustness.

Figure 3.10 presents the impact of gamma correction with a gamma value ranging from 0.4 to 2.0. Lin [19], Kozat [16], and Monga [24] achieve excellent robustness against gamma correction whereas Venkatesan [31] and Xie [36] have an impact on illumination changes caused by gamma correction. As regards Venkatesan [31], it becomes more sensitive to the gamma values of less than 0.6 while still having excellent robustness at the rest of its range. Compared to the other algorithms, Xie

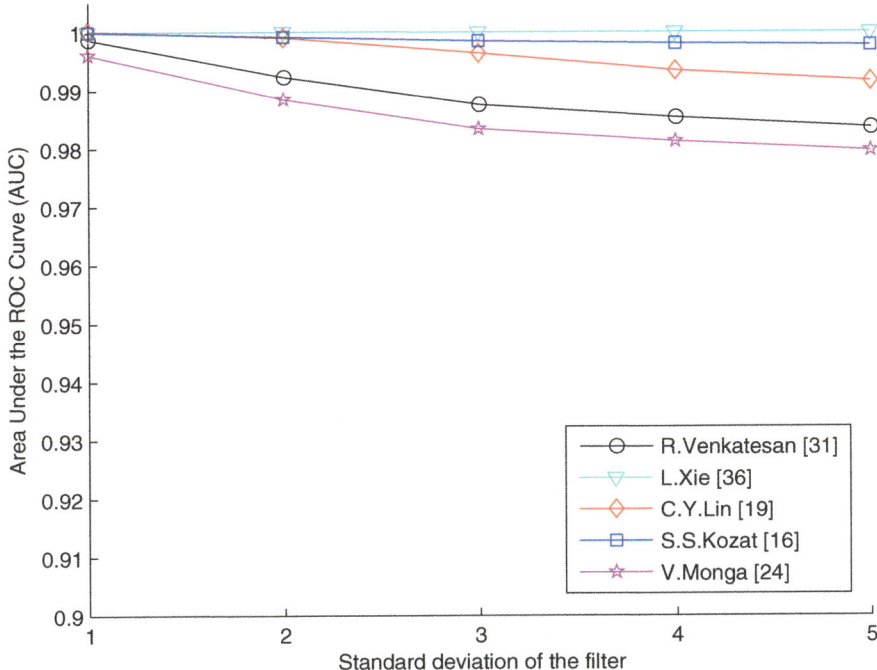

Fig. 3.8 Impact of Gaussian blurring

[36] is very sensitive to gamma correction since it is designed to utilize the MSB of 8 × 8 block's average intensity of an image as the invariant features.

To summarize, Kozat [16] demonstrates excellent robustness against all the content preserving modifications considered in this chapter. On the other hand, both Lin [19] and Monaga [24] are sensitive to AWGN while Xie [36] is sensitive to gamma correction. Furthermore, Venkatesan [31] shows a certain limitation on both AWGN and gamma correction.

3.6.2 Discriminability

As explained in Sects. 3.2 and 3.4, discriminability is one of the crucial requirements for content based image authentication using perceptual image hashing technique. In particular, malicious manipulations generally tend to be localized distortion, so that it is difficult to distinguish malicious manipulations since content preserving modifications globally distort the entire image like low-pass filtering and JPEG compression [4]. Accordingly, the experiment is designed to measure how well the selected algorithms can distinguish content changing manipulations from acceptable modifications which can preserve the content of an image. Concerning acceptable

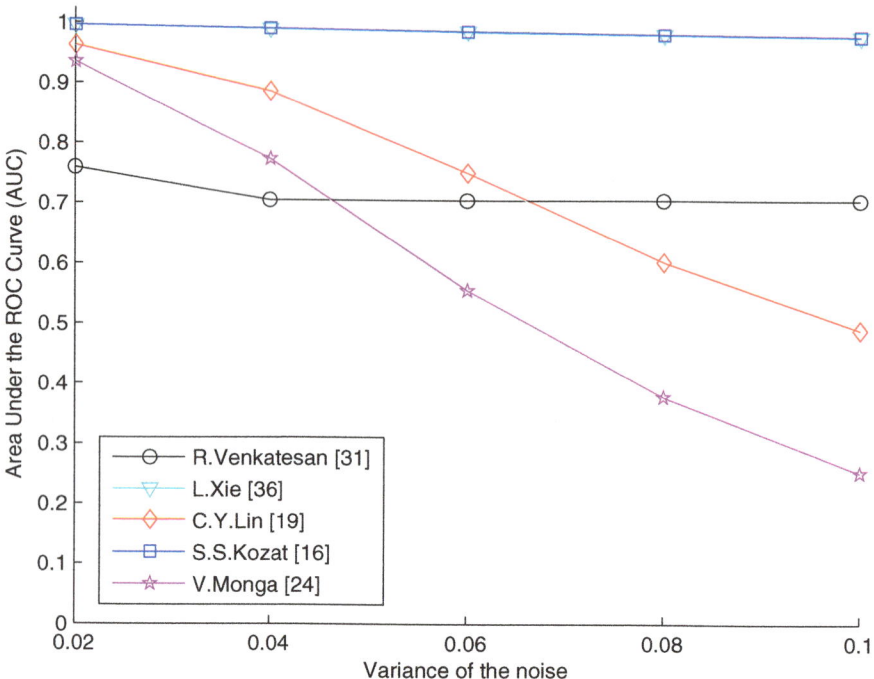

Fig. 3.9 Impact of additive Gaussian noise

modifications for this experiment, JPEG compression, scaling, and Gaussian blurring are used since all the selected algorithms have already demonstrated excellent robustness against those three modifications in Sect. 3.6.1. Figure 3.11 demonstrates the discriminability of the selected algorithms under the content changing manipulation as varying the size of a tampered region in the range between 16 × 16 and 128 × 128.

As shown in Fig. 3.11, the discriminability performance of all the selected algorithms significantly decreases as the size of a tampered region is decreased. Therefore, practically they cannot well discriminate the content changing manipulation from acceptable modifications in most cases. For instance, although Kozat [16] achieves 0.897 at the size of 128 × 128, the AUC value dramatically decreases and reaches 0.004 at the size of 16 × 16. Similarly, Venkatesan [31] and Xie [36] start with 0.800 and 0.792, and end with 0.2492 and 0.366. Those three algorithms manage to fairly distinguish the content changing manipulation at the size of larger than 96 × 96, but they can hardly differentiate the malicious tampering at the rest of its range. As for Lin [19] and Monaga [24], they are also not capable of discriminating the content changing manipulation when the size of a tampered region is smaller than 112 × 112 and 128 × 128 respectively. As a result, it is observed that all the selected algorithms cannot distinguish a small malicious manipulation from acceptable modifications in this experiment.

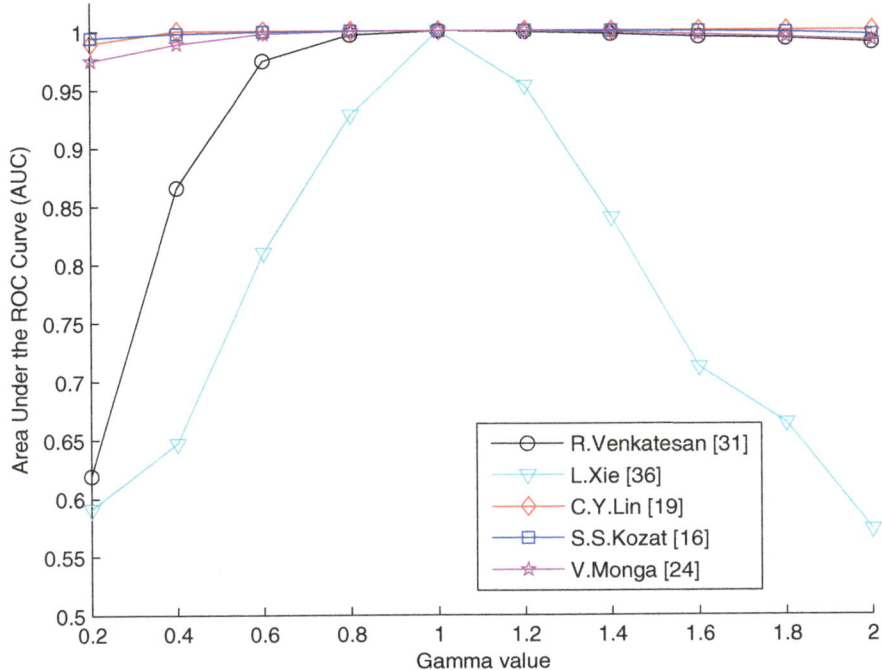

Fig. 3.10 Impact of gamma correction

3.6.3 Security

In general, a hash function itself is known to be insecure [26]. For instance, it may possible for an attacker to manipulate an image without changing a hash value based on his/her knowledge about the hash algorithm used. In this regard, many perceptual image hashing algorithms employ the randomness for their security aspect while others may simply encrypt a final hash value. To obtain the randomness, most perceptual image hashing algorithms utilize a secret key as the seed value of a random number generator, followed by randomly partitioning the image or the transformed image into several sub-blocks. The invariant features are extracted from each partitioned sub-block image, and then a final hash value is constructed by concatenating the results of corresponding sub-block images. Depending on the algorithm, random quantization may be employed as well in order to give a higher level of randomness to the final hash value. In such a way, a perceptual image hashing algorithm can achieve the security by making it difficult for an attacker to estimate or manipulate the final hash value without knowing the secret key even though the algorithm used is known.

To measure the randomness of the selected algorithms, the statistical irrelevance between a secret key and the corresponding hash value is investigated. In this experiment, the normalized Hamming distance between the hash value computed using an

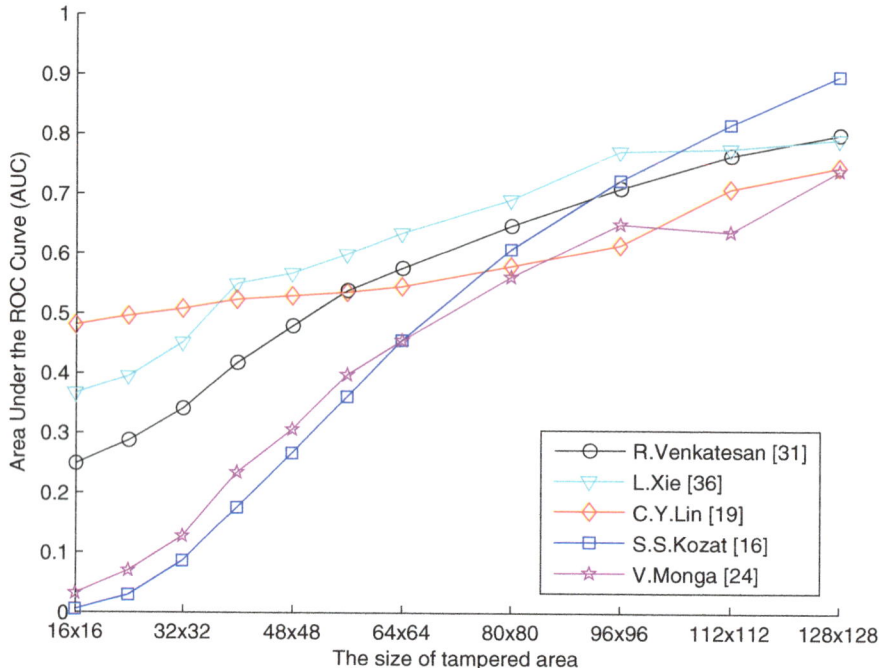

Fig. 3.11 Impact of content changing manipulation

initial secret key and other hash values generated using 1000 different secret keys is measured. As for the secret keys, the initial key is randomly selected using a random number generator, and then increased one by one to produce 1000 different secret keys. Figure 3.12 presents the experimental results for Venkatesan [31], Xie [36], and Kozat [16]. Note that, this experment is not applicable for Lin [19] and Monga [24] since they are designed to apply the existing digital signature algorithms on the final hash value instead supporting their own security mechanism.

As can be observed in Fig. 3.12, the normalized Hamming distance of Venkatesan [31], Xie [36], and Kozat [16] stay around 0.5—Mean distances are 0.428, 0.500, and 0.483 respectively—indicating the statistical irrelevance between the hash value and the secret key. Although three of them can achieve the sufficient randomness, it is potentially possible for an attacker to compromise a perceptual image hashing algorithm by exploiting the strong statistical and perceptual redundancy on the image content [9, 15, 20, 40].

3.7 Conclusion

Although a traditional data authentication scheme has provided a practical solution for legacy networks, image authentication for WMSNs is still a challenging problem since new requirements are emerged as mentioned in Sect. 3.2. As one of the

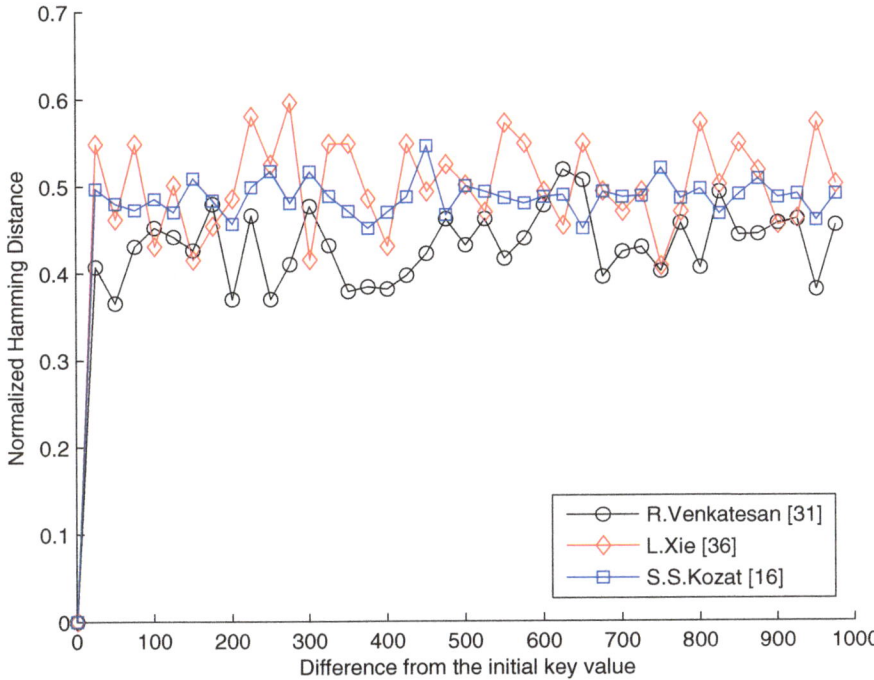

Fig. 3.12 Normalized hamming distance between the hash values computed with initial secret key and different secret keys

alternative solutions to cope with new requirements and challenges, content based image authentication using the perceptual image hashing technique has been introduced in this chapter. To investigate the feasibility of content based image authentication using the perceptual image hashing technique, several representative perceptual image hashing algorithms have been reviewed. Afterwards, five algorithms are selected to evaluate their performance with respect to robustness, discriminability, and security. Table 3.2 summarizes the results investigated in this chapter.

As for robustness, Kozat [16] which utilizes the dimensionality reduction technique demonstrates the best performance while the rest of algorithms have a certain limitation on AWGN and gamma correction. Unlike the robustness performance, any of five selected algorithms cannot properly discriminate content changing manipulations from content preserving modifications even though the size of a tamper region is large enough to effect perceptual changes in an image. From a security perspective, Xie [36] which employs the cryptographic primitive of AMAC possesses a good security property. Venkatesan [31] and Kozat [16] also achieve a moderate security by randomly partitioning an image into several sub-blocks and applying random quantization on the hash value.

Table 3.2 Comparison summary of the selected perceptual image hashing algorithms

Category/Algorithm			Comparison
Image statistics based approach	Venkatesan [31]	Feature	Image statistics (either averages or variances depending on the sub-band) of DWT coefficients calculated from the randomly partitioned rectangles of each sub-band in the wavelet decomposition of an image
		Robustness	Excellent robustness against JPEG compression, scaling, and Gaussian blurring, but having a certain limitation on AWGN and gamma correction
		Discriminability	Poor discriminability ($< 96 \times 96$)
		Security	Moderate security by randomly partitioning an image into several sub-blocks and applying random quantization on the hash value
			Weakness: Possible attacks that can modify an image without altering its image statistics
	Xie [36]	Feature	MSB of 8×8 image block average with image histogram transformation
		Robustness	Excellent robustness against JPEG compression, scaling, Gaussian blurring, and AWGN, but very sensitive to gamma correction. However, the original image should be modified by applying image histogram transformation before image transmission in order to obtain the good robustness
		Discriminability	Poor discriminability ($< 80 \times 80$)
		Security	Good security with a tamper detection capability by employing a cryptographic primitive AMAC and bit-level permutation
			Weakness: Possible attacks that can modify an image without altering its MSB of 8×8 image block average
Relation based approach	Lin [19]	Feature	Invariant relationship between DCT coefficients at the same position in a pair of 8×8 DCT blocks of an image
		Robustness	Excellent robustness against JPEG compression, scaling, Gaussian blurring and gamma correction, but having a certain limitation on AWGN

3 Perceptual Image Hashing Technique for Image Authentication in WMSNs

Table 3.2 (continued)

Category/Algorithm			Comparison
		Discriminability	Poor discriminability ($< 112 \times 112$)
		Security	Support a tamper detection capability, but no security mechanism on the hash algorithm
			Weakness: Possible attack that can produce a perceptually different image by slightly changing the low and middle frequency DCT coefficients of DCT blocks
Coarse image representation based approach	Kozat [16]	Feature	Strongest singular vectors of the SVD computed from the pseudo-randomly chosen semi-global regions of an image
		Robustness	Excellent robustness against JPEG compression, scaling, Gaussian blurring, AWGN, and gamma correction. However, the high computational complexity is required for iterating the SVD computation over each sub-block to capture the significant characteristics of an image
		Discriminability	Poor discriminability ($< 96 \times 96$)
		Security	Moderate security by randomly partitioning an image into several sub-blocks and applying random quantization on the hash value
			Weakness: False positive authentication with a high probability led by excellent robustness based on the dimensionality reduction technique
Low-level image representation based approach	Monga [24]	Feature	Salient feature points by using the end-stopped wavelet based feature detection algorithm
		Robustness	Excellent robustness against JPEG compression, scaling, Gaussian blurring, and gamma correction, but having a certain limitation on AWGN
		Discriminability	Poor discriminability ($< 128 \times 128$)
		Security	No security mechanism on the hash algorithm
			Weakness: Possible attacks that can add or remove a set of feature points for malicious manipulations while keeping the same set of the strongest feature points of an image due to the use of the limited number of feature points for the invariant features

As a result, content based image authentication using the perceptual image hashing technique still needs more improvements on robustness, discriminability and security to be an alternative solution for image authentication in WMSNs even if it has a huge technical potential. To enhance its performance regarding robustness and discriminability, the feature extraction method should be taken into account since the extracted features play the main role not only in obtaining robustness but also in detecting any malicious manipulations. However, there is a tradeoff between robustness and the discriminability since those requirements are in conflict with each other. As for security, there are still some possible attacks by exploiting the strong statistical and perceptual redundancy on the image content so that it is necessary to design a proper security mechanism according to the system requirements.

Acknowledgement This work was funded by the German Research Foundation (DFG) as part of the research training group GRK 1564 "Image New Modalities".

References

1. Ahmed F, Siyal MY. A secure and robust DCT-based hashing scheme for image authentication. IEEE Int. Conf. on Commun. Syst. 2006; pp. 1–6.
2. Akyildiz IF, Melodia T. Chowdhury KR. A survey on wireless multimedia sensor networks. J Comput Netw. 2007;59:921–60.
3. Boncelet C. The NTMAC for authentication of noisy messages. J IEEE Trans Inf Forensics Secur. 2006;1:35–42.
4. Chang E-C, Kankanhalli MS, Guan X, Huang Z, Wu Y. Robust image authentication using content based compression. J Multimed Syst. 2003;9:121–30.
5. Chen C-C, Lin C-S. Toward a robust image authentication method surviving JPEG lossy compression. J Inf Sci Eng. 2007;23:511–24.
6. Dawood MS, Ahila L, Sadasivam S. Image compression in wireless sensor networks—a survey. J Appl Inf Syst. 2012;1:11–5.
7. Graveman R, Fu K. Approximate message authentication codes. Proc. 3rd Annu. Fedlab Symp. on Adv. Telecommun./Inf. Distrib. 1999; vol. 1.
8. Guerrero-Zapata M, Zilan R, Barcel-Ordinas J, Bicakci K, Tavli B. The future of security in wireless multimedia sensor networks. J Telecommun Syst. 2010;45:77–91.
9. Han S-H, Chu C-H. Content-based image authentication: current status, issues, and challenges. J Inf Secur. 2010;9:19–32.
10. Haouzia A, Noumeir R. Methods for image authentication: a survey. J Multimed Tools Appl. 2008;39:1–46.
11. Harjito B, Han S, Potdar V, Chang E, Xie M. Secure communication in wireless multimedia sensor networks using watermarking. Proc. 4th IEEE Int. Conf. on Digit. Ecosyst. and Technol. 2010; pp. 640–5.
12. Harjito B, Potdar V, Singh J. Watermarking technique for wireless multimedia sensor networks: a state of the art. Proc. CUBE Int. Inf. Technol. Conf. 2012; pp. 832–40.
13. ISO/IEC 9797: Information technology—security techniques—message authentication Codes (MACs); 2011.
14. Jin Y, Vural S, Gluhak A, Moessner K. Dynamic task allocation in multi-hop multimedia wireless sensor networks with low mobility. J Sens. 2013;13:13998–4028.
15. Khelifi F, Jiang J. Analysis of the security of perceptual image hashing based on non-negative matrix factorization. IEEE Signal Process Lett. 2010;17:43–6.

16. Kozat SS, Venkatesan R, Mihcak MK. Robust perceptual image hashing via matrix invariances. Proc. IEEE Int. Conf. on Image Process. 2004; pp. 3443–6.
17. Kundur D, Luh W, Okorafor UN, Zourntos T. Security and privacy for distributed multimedia sensor networks. Proc IEEE. 2008;96:112–30.
18. Lin C-Y, Chang S-F. Semi-fragile watermarking for authenticating JPEG visual content. SPIE Secur. and Watermarking of Multimed. Contents II EI'00, SanJose, CA; 2000.
19. Lin C-Y, Chang S-F. A robust image authentication method distinguishing JPEG compression from malicious manipulation. J IEEE Trans Circuits Syst Video Technol. 2001;11:153–68.
20. Mao Y, Wu M. Unicity distance of robust image hashing. J IEEE Trans Inf Forensics Secur. 2007;2:462–7.
21. Masood H, Haider U, Ur-Rehman S, Khosa I. Secure communication in WMSN. Proc. Int. Conf. on Inf. Netw. and Autom. 2010; pp. 37–42.
22. Mihcak MK, Venkatesan R. New iterative geometric methods for robust perceptual image hashing. In: Revised Papers from the ACM CCS-8 Workshop on Security and Privacy in DRM 01, Tomas Sander, editor. London: Springer-Verlag; 2001.
23. Monga V, Evans BL. Perceptual image hashing via feature points: performance evaluation and tradeoffs. J IEEE Trans Image Process. 2006;15:3452–65.
24. Monga V, Vats D, Evans BL. Image authentication under geometric attacks via structure matching. IEEE Int. Conf. on Multimed. and Expo. 2005; pp. 229–32.
25. NIST: Digital Signature Standard (DSS). Federal Information Processing Standards Publication (FIPS PUB) 186-4; 2013.
26. Paar C, Pelzl J, Preneel B. Understanding cryptography: a textbook for students and practitioners. Berlin: Springer; 2010.
27. Perrig A, Szewczyk R, Tygar JD, Wen V, Culler DE. SPINS: security protocols for sensor networks. J Wirel Netw. 2002;8:521–34.
28. Queluz MP. Towards robust, content based techniques for image authentication. IEEE Int. Workshop on Multimed. Signal Process. 1998; pp. 297–524.
29. Shu L, Chen Y, Zhou Z. Watermarking technologies in wireless multimedia sensor networks. IEEE COMSOC MMTC E-Letter. 2012; pp. 24–7.
30. Ur-Rehman O, Zivic N. Fuzzy authentication algorithm with applications to error localization and correction of images. J WSEAS Trans Syst. 2013;12:371–83.
31. Venkatesan R, Koon S-M, Jakubowski MH, Moulin P. Robust image hashing. IEEE Int. Conf. on Image Process. 2000; pp. 664–6.
32. Vogt H. Integrity preservation for communication in sensor networks. Technical Report 434, ETH Zrich, Institute for Pervasive Computing; February 2004.
33. Walters JP, Liang Z, Shi W, Chaudhary V. Wireless sensor network security: a survey. In: Xiao Y, editor. Security in distributed, grid, and pervasive computing. Boston: Auerbach.
34. Wang H. Communication-resource-aware adaptive watermarking for multimedia authentication in wireless multimedia sensor networks. J Supercomput. 2013;64:883–97.
35. Wang H, Peng D, Wang W, Sharif H, Chen H-H. Energy-aware adaptive watermarking for real-time image delivery in wireless sensor networks. Proc. IEEE Int. Conf. on Commun. 2008; pp. 1479–83.
36. Xie L, Arce GR, Graveman RF. Approximate image message authentication codes. J IEEE Trans Multimed. 2001;3:242–52.
37. Zeng W, Yu H, Lin C-Y. Multimedia security technologies for digital rights management. Orlando: Academic Press, Inc.; 2006.
38. Zhang W, Liu Y, Das SK, De P. Secure data aggregation in wireless sensor networks: a watermark based authentication supportive approach. J Pervasive Mob Comput. 2008;4:658–80.
39. Zhou H, Guan X, Wu C. Reliable transport with memory consideration in wireless sensor networks. IEEE Int. Conf. on Commun. 2008; pp. 2819–24.
40. Zhu G, Huang J, Kwong S, Yang J. A study on the randomness measure of image hashing. J IEEE Trans Inf Forensics Secur. 2009;4:928–32.

Chapter 4
A Review of Approximate Message Authentication Codes

S. Amir Hossein Tabatabaei and Nataša Živić

4.1 Introduction

Fuzzy or approximate message authentication codes (AMACs) are cryptographic primitives which are not sensitive to minor changes in the received message. The acceptable changes in a message are reflected by the recognizable modifications in its authentication tag in such a scenario [1]. This essentially means that the authentication tag of an approximately-equal message to the original one should pass the verification test. These primitives have been explored significantly for the last two decades. The main applications arise in multimedia (image, video, etc.) and biometric authentication fields where the non-binary data might be tolerant against some content-preserving modifications caused by channel noise or partial faults in sensor reading. There exist valuable resources of the research works for these two main application fields, e.g., [2–8] based on the different classifications. When the message is not in a binary or a symbolic form, the different approaches for authentication are followed for feature vector extraction. An image feature vector can be extracted from block intensity histograms [9], edges [10], statistical features of the image [11], discrete cosine tranform (DCT), discrete wavelet transform (DWT) or Fourier transform (FT) coefficients [12–14]. When a message is binary, extracting a random substring after permutation is the main formal method for feature vector extraction in this case [14].

This chapter mainly considers a brief literature study on fuzzy message authentication codes, general concepts and notations and, accuracy and security requirements. Then, some improvements on the existing schemes are given and finally some applications in image authentication will conclude the chapter. We unify using the acronym AMAC for approximate message authentication codes which also point fuzzy message authentication codes to put the readers at ease.

S. A. H. Tabatabaei (✉) · N. Živić
Chair for Data Communications Systems, University of Siegen, 57076 Siegen, Germany
e-mail: amir.tabatabaei@uni-siegen.de

N. Živić
e-mail: natasa.zivic@uni-siegen.de

© Springer International Publishing Switzerland 2015
N. Živić (ed.), *Robust Image Authentication in the Presence of Noise*,
DOI 10.1007/978-3-319-13156-6_4

4.1.1 Definitions and Notations

Here, we introduce the required concepts and used notations which are essential for better explanation and analysis of the schemes. Unlike standard message authentication codes (MACs), in AMACs the acceptable changes in a message are reflected by the recognizable modifications in its authentication tag [1]. To measure the modification and also evaluate the authenticity, a distance function d and a threshold value δ are required. They are used to introduce approximate authentication by verifying the correctness of the received message having up to the acceptable differences from the original message [15]:

Definition 1 Let $\mathcal{M}$ be a finite metric space. A distance function on $\mathcal{M}$ is a function d such that $d : \mathcal{M} \times \mathcal{M} \rightarrow \mathcal{R}$ and for $\forall x, x' \in \mathcal{M}$, d satisfies the followings:

1. $d(x, x) = 0$
2. $d(x, x') = d(x', x) \geq 0$

Several different definitions of the AMAC exist in the literature [1, 14]. Let $\mathcal{K}$ be the key set, $\mathcal{M}$ be the message set, and $\mathcal{T}$ be the tag set. The following definition which is a represented form of the AMAC definitions appearing in [1, 14, 15] addresses all requirements for an approximate message authentication code.

Definition 2 A (d, p, δ)-approximate correct and $(d, \gamma, t, q, \epsilon)$-approximate secure message authentication code is a tuple $(\mathcal{K}, \mathcal{M}, \mathcal{T}, d, Kg, Tag, VF)$ such that the following requirements are satisfied:

1. d, Kg, Tag and VF are four polynomial-time algorithms.
2. $Kg(., d) : \{0, 1\}^l \rightarrow \{0, 1\}^l$ is a key generator algorithm which generates an l-bit key on l-bit input.
3. Tag : $\mathcal{K} \times \mathcal{M} \rightarrow \mathcal{T}$ is the tag generator algorithm which generates an authentication tag on an input message and a key.
4. VF : $\mathcal{K} \times \mathcal{M} \times \mathcal{T} \rightarrow \{0, 1\}$ is a verification algorithm such that:

$$\forall k \in \mathcal{K}, \forall m \in \mathcal{M}, Tag(k, m) = t \iff VF(k, m, t) = 1.$$

5. "(d, p, δ)-Approximate correctness" [1]: $\forall m, m' \in \mathcal{M}, \forall k \in \mathcal{K}, Prob\{VF(k, m', t) = 1 \mid d(m, m') \leq \delta, Tag(k, m) = t\} \geq p$.
6. "$(d, \gamma, t, q, \epsilon)$-Approximate security" [1]: $\forall (m_i, t_i) \in \mathcal{M} \times \mathcal{T}, i = 1...q : Tag(k, m_i) = t_i, Prob\{VF(k, m, t) = 1 \mid d(m, m_i) \geq \gamma\} \leq \epsilon$.

Some equivalent terms have been used in the literature to address the above properties in an AMAC scenario like robustness and sensitivity [6, 16]. The last two properties address the unique features of an AMAC. In fact (d, p, δ)-approximate correctness introduces robustness into the mechanism while $(d, \gamma, t, q, \epsilon)$-approximate security indicates its sensitivity. Also some additional conditions may be considered for security analysis, like the following from [1]:

Definition 3 A (d, p, δ)-approximate correct and $(d, \gamma, t, q, \epsilon)$-approximate secure message authentication code is weak preimage resistant if an adversary with sufficient number of valid pairs $(m_i, \mathsf{Tag}(k, m_i))$ has a negligible chance to find a message m' which passes the verification test, i.e., $\mathsf{VF}(k, m', t_i) = 1$.

If the additional assumption of accessibility of adversary to the secret key k is given in the above definition, the latter property is called "strong preimage-resistance" [1]. It is worth mentioning that (m, t) is called a valid pair when $\mathsf{VF}(k, m, t) = 1$ for a received message m [1]. The security issues for an AMAC is not limited to the above definitions. It should be computationally difficult to recover the secret key with sufficient number of valid message-tag pairs.

4.2 Dedicated AMAC Schemes

In this section, the main dedicated AMAC mechanisms are introduced and criticized. From designing point of view, the AMACs either can engage standard cryptographic primitives as their building blocks or use some non-cryptographic robust functions.

4.2.1 Majority-Based AMACs

Graveman and Fu were the first authors who proposed a generic AMAC scheme for binary messages in 1999 [17]. Their design was quite simple and creative. The corresponding tag generation algorithm works as follows: the binary message is firstly zero-padded if required. Then it is reshaped by splitting into blocks and then permuted. The permuted message is randomized by XORing with a generated random number sequence based on a shared secret key between the sender and the receiver. The final AMAC tag is a probabilistic checksum calculated by a MAJORITY selection round. In an extension, Ge et al. [18] generalized the proposed AMAC scheme to the non-binary alphabets which is called N-ary AMAC ($N \geq 2$). The application of the generalized scheme has been shown for image authentication purposes as well. This application will be addressed in the last section of this chapter. When $N = 2$ (binary alphabet) in the N-ary AMAC, it is converted to the initial AMAC obviously. Thus, here we just explain the specification of the N-ary AMAC which is the generalized form of the initial scheme. The tag generation algorithm of the N-ary AMAC generates the message probabilistic checksum using the MAJORITY selection function like in the initial AMAC scheme [17]. A pseudo random number generator and a random permutation function is used for randomizing the N-ary elements based on the shared secret key. The algorithm specification comes in the following.

Algorithm 1 N-ary AMAC tag generation [18]

Input: message $x \in \mathcal{M} \subseteq \mathbb{Z}_N^m$ of length m, shared secret key k generated by $\text{Kg}(.,d)$, matrix size R, L

1: **procedure** $\text{N} - \text{aryAMAC} - \text{Tag}(k,x)$
2:    x'=zero padding on x ▷ The message x is zero padded to generate x', an $R \times L$-word sequence if required.
3:    Z=reshaping of x'.   ▷ x' is reshaped form of an $R \times L$ matrix Z
4:    $Q = \text{Permute}(Z,k)$ ▷ $\text{Permute}(.,k)$ is a permutation algorithm.
5:    **for** i=1: R **do**   ▷ R and L are input parameters for a message block defined according to recommendations specified in [18].
6:       **for** j=1: L **do**
7:          $Q_{ij} = Q_{ij} \oplus r_{ij}$ ▷ r_{ij}'s are the generated random numbers by $\text{PRNG}(k)$.
8:       **end for**
9:    **end for**
10:    **for** $j = 1, L$ **do**
11:       $t_j = \text{MAJORITY}(Q_j)$ ▷ MAJORITY selection is performed on each column of the generated matrix Q.
12:    **end for**
13:    $T = t_1, ..., t_L$
14:    **return** T
15: **end procedure**

Algorithm 1 calculates the tag of a message of length $R \times L$ elements from the message set $\mathbb{Z}_N^m$. For messages with arbitrary length, the message is split into smaller blocks of length $R \times L$ after zero-padding (if required). Then the tag generation algorithm specified above in Algorithm 1 is performed on each block. The final tag is generated by applying the stepts 5-12 of Algorithm 1 on the derived intermediate tag array of the message blocks. The final tag is a probabilistic MAJORITY-based checksum consisting of L N-ary elements from $\mathbb{Z}_N$. The verification algorithm of N-ary AMAC measures the distance between the reference authentication tag and the one calculated from the received message using Hamming distance function. The verification is successful and the message is declared as authentic if the calculated distance between the reference tag and the recalculated one (from the received message) is not beyond the predefined threshold.

4.2.1.1 Shifting Attack

The generalized N-ary AMAC has a very simple structure based on the pioneered scheme. However, it has a weak performance and a proven security flaw which will be mentioned. The provided robustness is based on the noise stability of the MAJORITY function [19]. This property also increases inaccuracy of the mechanism in the verification algorithm, i.e., the N-ary tag generation algorithm can produce very similar tags for non-enough close messages which cancels the approximate security and approximate correctness properties supposed to be provided by an AMAC scenario. Nevertheless, it remained secure for more than a decade and no generic deterministic attack has been proposed against it till late 2011 [20]. Based on an implementation observation for $N = 2$, an AMAC tag of a complemented message can be extracted just by complementing the tag of the original message. This correlation weakness could appear later as a special case of a generic attack on N-ary AMAC in [16]. Tonien et al. [16] presented the first (and the only) substitution passive attack based on just one observed authenticated message-tag pair. The attack uses two potential weaknesses of the original AMAC located in its padding scheme and the MAJORITY selection parts. This linear distance preserving property of the MAJORITY selection function was a trigger for proposing a generic forgery attack on this AMAC. This hidden property is called shift-invariance property which indicates the prefect linearity characteristic of the MAJORITY selection function. Shift-invariance property indicates that for any message x of length L with elements in $\mathbb{Z}_N$ and any constant $c \in \mathbb{Z}_N$, $\text{MAJORITY}(x+c) = \text{MAJORITY}(x) + c$. The addition is computed in modulus N, obviously. This linear property of the MAJORITY function along with the linear property of the other parts of the tag generation algorithm (including padding step) results in $\text{N}-\text{aryAMAC}-\text{Tag}(k, x+c) = \text{N}-\text{aryAMAC}-\text{Tag}(k, x) + c$. The attack is called shifting attack [16] due to the latter equation. The attacker who launches a shifting attack, has observed an authenticated message-tag pair (x, T). Then the attacker chooses an arbitrary constant value c from the corresponding message set. The $(m+c, T+c)$ will be a new authentic message-tag pair. The authenticity of the new pair can easily be proven from the latter shift-invariance property.

4.2.1.2 Security Enhancement

To enhance the security against shifting attack and to remove this serious weakness from the AMAC scheme, some suggestions have been given by the authors in [16]. According to the proposed solution in [16], the old padding procedure is substituted by the SHA-1 padding scheme and the nonlinearity is introduced to the randomization step. In the randomization step, two random bits r_{ij} and r'_{ij} are generated for each bit Q_{ij} and the linear operation $Q_{ij} = Q_{ij} \oplus r_{ij}$ is substituted by the new nonlinear operation $Q_{ij} = r_{ij} Q_{ij} \oplus r'_{ij}$. Also the ambiguity of the MAJORITY selection part is removed. The details can be studied in [16]. The proposed solution removes the vulnerability of the N-ary AMAC against shifting attack by introducing the nonlinearity into the corresponding tag generation algorithm and destroying the existing shift-invariance property.

In a recently proposed AMAC, a statistical feature of a random sequence together with a Boolean function is used to generate an authentication tag. The introduced AMAC scheme in [20] uses an ITE Boolean function, number of runs of ones and a MAJORITY selection function for a tag generation. The ITE Boolean function outputs a binary value for three input binary variables as $\text{ITE}(a, b, c) = ab \oplus ac \oplus c$. The cryptographic properties of the ITE have been analyzed in [21] in detail. It is mentioned that the special case of $N=2$ is considered here. A run of ones is a sequence of consecutive ones preceded and succeeded by zeros or by nothing [22]. The statistical distribution of the runs of ones in a binary sequence has been significantly useful in lots of important applications. The application domains contain a diverse range of fields like bioinformatic, computational biology, encoding, and compression [20, 23–25]. Also, the number of runs of ones and their maximum are considered as the statistical features of a random binary sequence which are used in pseudo-randomness tests for binary sequences [20, 26]. The robustness of the algorithm for similar messages is preserved. The algorithm specification for generating the intermediate tag is given by Algorithm 2 and is called Boolean function-based AMAC or simply BAMAC. The final tag size of the proposed scheme is relatively larger than the similar approach [20]: To calculate the final tag for an arbitrary size message, after dividing the message into M equal size blocks and applying Algorithm 2, the generated integer matrix is represented in the binary form and is subjected to the light version of Algorithm 2 where the run calculation is substituted by a MAJORITY selection function to generate the final tag.

Algorithm 2 BAMAC tag generation [20]

Input: message $x \in \mathcal{M} \subseteq \mathbb{Z}_2^m$ of length m, shared secret key k generated by $\mathsf{Kg}(.,d)$, matrix size R, L

1: **procedure** BAMAC − Tag(k, x)
2:      Divide the message into M equal size blocks of length $R \times L$
3:      **for** $i = 1 : M$ **do**
4:          $x' = \mathsf{SHA1} - \mathsf{Padding}(x)$ ▷ The message x is padded according to the padding algorithm of SHA1 hash function.
5:          Z = reshaping on x'.   ▷ x' is reshaped to the form of an $R \times L$ matrix Z
6:          $Q = \mathsf{Permute}(Z, k)$ ▷ $\mathsf{Permute}(., k)$ is a permutation algorithm.
7:          **for** j=1: R **do**
8:              **for** t=1: L **do**
9:                  $Q_{jt} = \mathsf{ITE}(Q_{jt}, r_{jt}, s_{jt})$ ▷ r_{jt}'s and s_{jt}'s are generated random numbers by PRNG(k).
10:              **end for**   ▷ $\mathsf{ITE}(x_1, x_2, x_3) = x_1 x_2 \oplus x_1 x_3 \oplus x_3$
11:          **end for**
12:          **for** $j = 1, L$ **do**
13:              Calculate the number of runs of ones in each column of matrix Q and assign it to t_{ij}.
14:          **end for**
15:      **end for**
16:      **return** $[t_{ij}]$ ▷ The intermediate tag value is an integer matrix of size $N \times L$
        ($0 \leq t_{ij} \leq \lfloor \frac{m+1}{2} \rfloor$).
17: **end procedure**

4.2.1.3 Analysis of the Majority-Based AMACs

Here, the analysis based on the first introduced approach is given. The extension to the generalized schemes are straight forward.

Let the sender and the receiver agree on the AMAC scenario. Here, we calculate the changes which are reflected into the generated tags corresponding to the modified messages. Suppose that two authentic pairs (m, t) and (m', t') are given such that $HD(m, m') = d$. We calculate the expected Hamming distance between the authentication tags t and t'. Following the above notation, suppose that the message is padded and split into P blocks each of size $R \times L$-bit. It is known that due to the randomness of the blocks (the randomness is because of XORing with a random binary block) the distribution of the Hamming distance among each column follows a binomial distribution. Let D denote the random variable of differences between the messages in each column then its probability is computed as follows.

$$P_D(l) = \frac{\binom{R}{l}}{2^R} \qquad (4.1)$$

Now, the distribution function for the number of columns with k bits differences among of all blocks is calculated. Let f_K denote the corresponding distribution function, then it is calculated according to a classical occupancy problem [17, 27]:

$$f_K(k) = \frac{\binom{PL+d-k-2}{d-k}}{\binom{PL+d-1}{d}} \qquad (4.2)$$

Using Eq. 4.1, we can calculate the probability P_L of changing the column MAJORITY function with e bits differences in the corresponding column as follows.

$$P_{R,e} = 2 \sum_{i=1}^{e} \sum_{j=0}^{\lfloor \frac{e-i}{2} \rfloor} \frac{\binom{\lceil \frac{R}{2} \rceil + i - 1}{e-j} \binom{R - \lceil \frac{R}{2} \rceil - i + 1}{j}}{\binom{R}{e}} P_D(\lceil \frac{R}{2} \rceil + i - 1) \qquad (4.3)$$

The expected value of differences between the MAJORITY selection row of each block is calculated as $E_L = \sum_K L P_{R,k} f_K(k)$. Finally, by reapplying the classical occupancy problem the probability distribution of the E_L among all P MAJORITY selection block arrays is calculated as below:

$$f_{E_L,d'}(e_L) = \frac{\binom{P+d'-e_L-2}{d-e_L}}{\binom{P+d'-1}{d'}} \qquad (4.4)$$

where d' indicates the differences between two corresponding rows of the MAJORITY selection block arrays. The expected difference values between two final tags t and t' considering $HD(m, m') = d$ is calculated as follows.

$$E_{HD(t,t')} = \sum_{i=1}^{d} L f_{E_L,i}(e_L) P_{R,i} \qquad (4.5)$$

According to the Eq. 4.5 one can find the expected Hamming weight between two authentication tags corresponding to different messages which gives an indication of the threshold setting to verify the authenticity. To keep the scheme reliable for big size messages and avoid security vulnerabilities it is recommended [16] to use normalized Hamming distance (NHD) instead. The normalized Hamming distance is calculated by dividing the computed Hamming distance on the message or the tag length. Using the derived Eq. 4.5 the user can set a threshold value for a false acceptance and a false rejection according to the acceptable changes in the original message. The detailed security analysis based on the false acceptance and false rejection events are given in [17, 16].

The Fig. 4.1 shows the observed distance between the BAMAC tags versus distance between the input messages for $N = 33$, $R = 127$ and $L = 256, 512$. The comparison between the sensitivity of the initial AMAC, repaired one [16] and the proposed AMAC in [20] is depicted in Fig. 4.2.

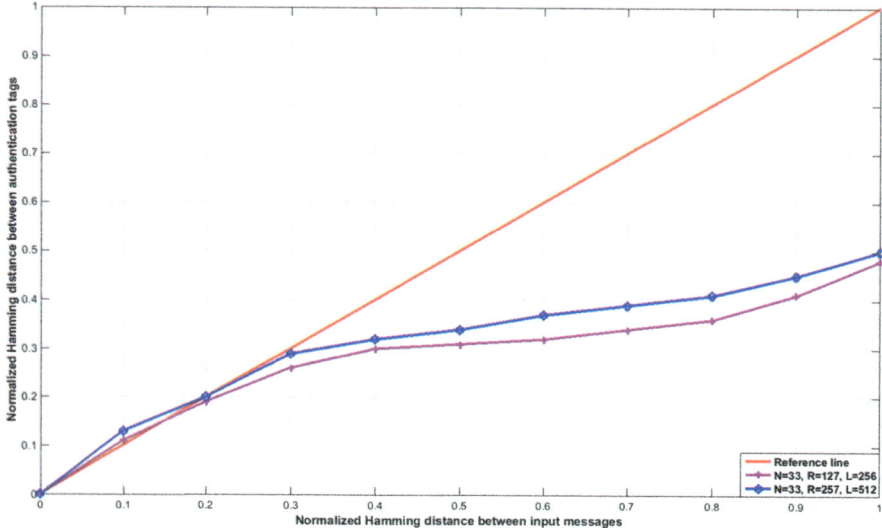

Fig. 4.1 Sensitivity of the BAMAC tag against input changes [20]

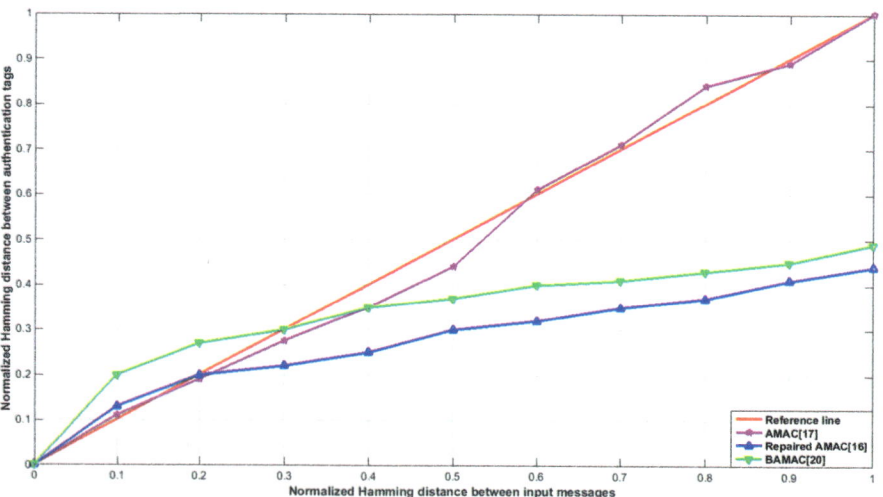

Fig. 4.2 Comparison between the nonlinearity behavior of three AMACs (the proposed algorithm refers to the BMAC tag generation scheme) [20]

4.2.2 *Noise Tolerant Message Authentication Codes (NTMACs)*

Noise tolerant message authentication code which is briefly called NTMAC is an AMAC scheme which has been proposed by Boncelet [28] in 2006. Unlike the latter approach, the NTMAC uses cryptographic MAC in its structure. The tag generation algorithm of the NTMAC first divides the message into blocks repeatedly and then calculates the truncated ordinary MAC of each block called subMAC. The final tag is the concatenation of these subMACs. The main advantage of this approach is ability to locate the erroneous message blocks. This point can be used further to avoid resubmitting the whole message when the authentication fails. The specification of the algorithm is given below in Algorithm 3.

Algorithm 3 NTMAC tag generation [28]

Input: message $x \in \mathcal{M} \subseteq \mathbb{Z}_2^m$, shared secret key k generated by Kg(.,d), number of blocks n, number of partitions t
1: **procedure** NTMAC − Tag(k,x)
2:     $\{x_1,...,x_t\}$ = Partition(x,k) ▷ x_i is a message partition which contains disjoint message blocks $x_{i1},...,x_{in}$ such that $\bigcup_{j=1}^{n} x_{ij} = x$
3:     **for** i=1:t **do**
4:        **for** j=1: n **do**
5:           $H_{ij} = SUBMAC_k(x_{ij})$ ▷ H_{ij}s are the truncated MACs of blocks x_{ij}s.
6:        **end for**
7:     **end for**
8:     $T = H_{11}||...||H_{tn}$
9:     **return** T
10: **end procedure**

The tag generation algorithm of the NTMAC computes the MAC of each message block which imposes high computational cost for implementation. However, it is the first generic fuzzy message authentication with error localization property which can be used to look for errors in non-authentic blocks or to resubmit them. Some extensions have been made based on the NTMAC. To decrease the overall collision rate, the cyclic redundancy check (CRC) is used following with a block cipher encryption algorithm in CRC-NTMAC [29] to provide the security(see Algorithm 4). Although the probability of finding a collision is reduced generally in the latter scheme, the choice of CRC is very important and a wrong selection of CRC makes it extremely vulnerable against key recovery attack.

Algorithm 4 CRC-NTMAC tag generation [29]

Input: message $x \in \mathcal{M} \subseteq \mathbb{Z}_2^m$, shared secret keys k_1, k_2 generated by Kg(.,d), number of blocks n, number of partitions t
1: **procedure** CRC − NTMAC − Tag (k_1, k_2, x)
2:     $\{x_1,...,x_t\}$ = Partition(x,k_1) ▷ x_i is a message partition which contains disjoint message blocks $x_{i1},...,x_{in}$ such that $\bigcup_{j=1}^{n} x_{ij} = x$
3:     **for** i=1:t **do**
4:        **for** j=1: n **do**
5:           $H_{ij} = CRC(x_{ij})$ ▷ H_{ij}s are CRCs of blocks x_{ij}s.
6:        **end for**
7:     **end for**
8:     $T = \mathrm{E}(H_{11}||...||H_{tn}, k_2)$ ▷ $\mathrm{E}(.,k_2)$ is a an encryption algorithm of a block cipher.
9:     **return** T
10: **end procedure**

In another extension of the NTMAC, the CRC is replaced by the BCH code as a class of the CRC codes. This improves the performance in the sense of better estimation of errors number [30].

The NTMAC scheme has been used in an image authentication scheme later on. The NTMACs are using a classic MAC as their building blocks so they convey their security properties but at the cost of computational load. To tackle with this drawback, one approach is to assign some weight coefficients to the message parts to increase the efficiency. Also, these schemes provide partial error correction capability which was not given in the previous schemes.

Error correcting NTMAC (EC-NTMAC) and error correcting weighted NTMAC (EC-WNTMAC) are two authentication and verification mechanisms which have been proposed in [31] based on the NTMAC. The first approach provides error detection and correction and the second approach assigns some weights to the message blocks according to their importance level. So, the high importance message blocks are supported by longer authentication tag in comparison to the message blocks with lower importance. The EC-NTMAC tag generation algorithm is given below (Algorithm 5).

Algorithm 5 EC-WNTMAC tag generation [31]

Input: message $x \in \mathcal{M} \subseteq \mathbb{Z}_2^m$, shared secret key k generated by $Kg(.,d)$, number of blocks n, block weights $w_1,...,w_n$, number of partitions t

1: **procedure** $EC-WNTMAC-Tag(k,x)$
2:     $\{x_1,...,x_t\} = Partition(x,k)$ ▷ x_i is a message partition which contains disjoint message blocks $x_{i1},...,x_{in}$ such that $\bigcup_{j=1}^{n} x_{ij} = x$
3:     **for** i=1:t **do**
4:        **for** j=1:n **do**
5:           $H_{ij} = SUBMAC_k(x_{ij})$ ▷ H_{ij}'s are MAC of blocks x_{ij}s truncated according to their weights w_j.
6:        **end for**
7:     **end for**
8:     $T = H_{11}||...||H_{tn}$
9:     **return** T
10: **end procedure**

The error correction capability is provided by soft information of the channel output. It uses bit-flipping of the bits with lowest absolute reliability values ($LLRs$) to correct the suspicious message blocks [31]. The specification of the verification algorithm of the EC-WNTMAC follows in Algorithm 6.

Algorithm 6 EC-WNTMAC verification [31]

Input: received message $x' \in \mathcal{M} \subseteq \mathbb{Z}_2^m$, received tag T, shared secret key k generated by $Kg(.,d)$, number of blocks n, block weights $w_1,...,w_n$, LLR values, iteration number $itr_1,...,itr_n$ for each block, number of partitions t

1: **procedure** EC − WNTMAC − VF(k, x', T)
2:     $\{x'_1,...,x'_t\}$ = Partition(x', k) ▷ x'_i is a message partition which contains disjoint message blocks $x'_{i1},...,x'_{in}$ such that $\bigcup_{j=1}^{n} x'_{ij} = x'$
3:     set $T = H_{11}||...||H_{tn}$
4:     **for** i=1:t **do**
5:         **for** j=1:n **do**
6:             $H'_{ij} = SUBMAC_k(x'_{ij})$
7:             **if** $H_{ij} = H'_{ij}$ **then**
8:                 a_{ij}=1   ▷ a_{ij} is set to 1 for an authentic block x_{ij} otherwise 0.
9:             **else**
10:                **if** Number of iteration is not achieved, **then**
11:                    Arrange the bits of x'_{ij} in the increasing order of their $|LLR|$ values and flip the next combination of least reliable bits of x_{ij} and **go to** 6.
12:                **end if**
13:            **end if**
14:        **end for**
15:    **end for**
16:    **return** $a_{11},...,a_{in}$
17: **end procedure**

The following section gives the performance and security analysis of the NTMAC based schemes.

4.2.2.1 Analysis of the NTMAC-Based AMACs

Here, the analysis of the AMACs which use the NTMAC core is given. To have a generic analysis, we consider the analysis of the EC-WNTMAC [31] which also includes the simpler versions of the NTMAC. To formalize the concept concerning Algorithm 6 the basic random binary variables $B_{i,j}$, $F_{i,j}$ and their cumulative versions are defined as follows to indicate the erroneous message and MAC blocks respectively [28, 31].

$$B_{ij} = \begin{cases} 1, & x_{ij} \neq x'_{ij} \\ 0, & x_{ij} = x'_{ij} \end{cases}, \quad B_i = \sum_{j=1}^{n} B_{ij}, \qquad (4.6)$$

$$B = \sum_{i=1}^{t} B_i$$

$$F_{ij} = \begin{cases} 1, & H_{ij} \neq H'_{ij} \\ 0, & H_{ij} = H'_{ij} \end{cases}, \quad F_i = \sum_{j=1}^{n} F_{ij}, \qquad (4.7)$$

$$F = \sum_{i=1}^{t} F_i$$

Also, we suppose that E_{ij} and E indicate the number of bit errors in a message block and the whole message respectively. Using above setup and assuming that an ideal standard MAC is used to generate the MAC blocks we have: $Pr(F_{ij} = 1|B_{ij} = 1) = 1 - 2^{-|H_{ij}|}$. Let the received message contain η erroneous bits then, we estimate the number of erroneous message blocks using Maxwell–Boltzmann statistics [27] and the classical occupancy problem when the errors are distributed independently. In this regard, we have the following [31]:

$$Pr\{\text{each message-block contains at least one error}\}$$
$$= \frac{(-1)^i \sum_{i=0}^{n} (-1)^i \binom{n}{i}(n-i)^\eta}{n^\eta} \qquad (4.8)$$

The probability of false rejection of a message depends on the bit error rate (BER) of the communication channel. Let θ denote the BER, the probability of a false rejection is calculated as follows [31]:

$$P_{FR} = Pr(F_{ij} = 1|B_{ij} = 0) = \qquad (4.9)$$

$$1 - (1 - \theta)^{|H_{ij}|} \qquad (4.10)$$

A false acceptance event happens when the error bits in the message are not notified by the verification algorithm. This event reduces the security of the scheme and can be used by the attacker in several scenarios. The probability of a false acceptance can be denoted by $P_{FA} = Pr(F = 0|E = \eta)$. Using basic probability theory, P_{FA} is calculated as follows [31].

$$P_{FA} = \prod_{i=1}^{m} \sum_{j=0}^{n} Pr(F_i = 0|B_i = j).Pr(B_i = j|E = \eta) \qquad (4.11)$$

The components of the above equation are evaluated as follows [31]:

$$Pr(F_i = 0|B_i = j) = \frac{\sum_{k=1}^{\binom{n}{j}} 2^{-T_k}}{\binom{n}{j}} \qquad (4.12)$$

$$Pr(B_i = j|E = \eta) = \binom{n}{j} \sum_{t=0}^{j} \frac{(-1)^t \binom{j}{t}(j-t)^\eta}{n^\eta} \qquad (4.13)$$

T_k indicates the summation of j-th sub-MAC lengths (as they can have different lengths depending on the weights $w_1, \ldots, w_n$) out of n per each selection in all possible combinations. Combining the above equations together results in the expanded form for the P_{FA} as follows [31]:

$$P_{FA} = (\sum_{j=0}^{\eta} \sum_{k=1}^{\binom{n}{j}} 2^{-T_k} \sum_{t=0}^{j} \frac{(-1)^t \binom{j}{t}(j-t)^\eta}{n^\eta})^t \qquad (4.14)$$

The EC-WNTMAC scheme has an enhanced error localization ability which allows the corresponding verification algorithm to identify the erroneous block high likely. Also the expected number of errors are estimated more accurately in comparison to the basic NTMAC. Further information are given in [31]. The number of expected errors decreases by using NTMAC, CRC-NTMAC, and BCH-NTMAC schemes respectively. It is worth mentioning that the introduced error localization property in the NTMAC-based family of AMACs can reduce the security as well. An intelligent choice of the chosen plaintexts can be used to find the permutation used in the corresponding tag generation algorithms. To overcome this weakness, it is recommended to change the secret keys after reasonable usage times.

4.2.3 AMACs Based on Computational Security

Crescenzo et al. have introduced two rigorous AMAC schemes in [1]. The proposed schemes use the symmetric encryption algorithm, the target collision resistant (TCR) hash functions and the error-correcting code. The formal security analysis in the computational Security viewpoint shows that the proposed schemes do not have the security flaws of the previous generic schemes since the standard cryptographic primitives are used as their building blocks ingeniously [15].

In the first approach, which is a weak preimage resistant scheme [1], the authentication tag is created based on a symmetric encryption algorithm (E_k, D_k), a systematic error-correcting code (Enc, Dec) and a standard MAC, $MAC(k,.)$. E_k (D_k) indicates encryption (decryption) algorithm and Enc (Dec) indicates encoding (decoding) algorithm respectively. The error correction capability of (Enc, Dec) is δ. This scheme is a (d, p, δ)-approximate correct and (d, γ, t, q, ϵ)-approximate secure message authentication code with optimal parameters $p = 1$ and $\gamma = \delta + 1$ [1, 15]. The tag generation algorithm is given as follows in Algorithm 7.

Algorithm 7 AMAC$_1$ tag generation [1]

Input: shared secret keys k_1, k_2 generated by Kg(., d), m-bit message x, error correction capability δ

1: **procedure** AMAC$_1$ − Tag(k_1, k_2, x)
2:     $x \| pc = Enc(x, \delta)$ ▷ pc indicates the parity check bits.
3:     $t_1 = MAC(k_1, x)$
4:     $C = E_{k_2}(pc)$
5:     $tag = C \| t_1$
6:     $T = tag$
7:     **return** T
8: **end procedure**

The verification algorithm of AMAC$_1$ attempts to correct the decoded decrypted message by the error-correcting code if the number of detected errors are beyond its capability until the MAC matches (Algorithm 8) [1].

Algorithm 8 AMAC$_1$ verification [1]

Input: shared secret keys k_1, k_2 generated by Kg(.,d), m-bit received message x', error correction capability δ, reference tag T

1: **procedure** AMAC$_1$ − VF(k_1, k_2, x', T)
2:     $T = pc \| t_1$ ▷ pc and t_1 indicate the parity check part and the MAC part respectively.
3:     $pc = D_{k_2}(pc)$
4:     $X' = Dec(x' \| pc, \delta)$
5:     If number of detected errors in X' is beyond δ **return** 0 and stop.
6:     If $MAC(k_1, X') = t_1$ **return** 1 else **return** 0
7: **end procedure**

The proposed authentication scheme AMAC$_1$ meets weak pre-image resistance property and enables partial error correction. It does not tolerate any mismatch between the MACs. Also error localization property is not provided.

In the second proposed scheme in [1], a strong preimage resistant scheme, the authentication tag is generated based on the construction of probabilistic universal one-way hash functions [15, 32]. The corresponding authentication mechanism uses a pseudo-random function for extracting a random permutation and a random number sequence. In the description of the tag generation algorithm 9, $PRF_k(.)$ and π indicate a pseudo-random function and a random permutation respectively. Also the finite target collision resistant hash function is denoted by $UHF_u(.)$.

The output tag length of Algorithm 9 is $t_2.b$ bits where $|h_i| = b$ bits. The corresponding verification algorithm differs from standard MACs due to approximate correctness property. In this scheme, the verification algorithm checks for equality between the received and the computed hash-block sequences in a defined number of positions Tr. Unlike the first scheme, the AMAC$_2$ tolerates errors in the authentication tag up to a certain limit. Also error localization property can be achieved via an intelligent selections of the blocks. However, error correction is not possible in the current scheme. The specification of the corresponding verification algorithm is given in Algorithm 10.

Algorithm 9 AMAC$_2$ tag generation [1]

Input: shared secret key k generated by Kg(.,d), initial value IV, m-bit message x, block size c, number of blocks n, approximate correctness parameters p and δ, approximate security parameter γ

1: $t_1 = \lceil m/2c\delta \rceil$, $t_2 = \lceil -10\log(1-p) \rceil$
2: **procedure** AMAC$_2$ − Tag(k, x)
3:     $(u \| \pi \| \rho) = PRF_k(IV)$ ▷ $u \in \{0,1\}^k$, π is a permutation and ρ is a random selection function.
4:     $x_1 \| ... \| x_n \leftarrow \pi(x)$
5:     **for** $i = 1, t_2$ **do**
6:       $N_i^\rho \leftarrow x_{i_1} \| ... \| x_{i_{t_1}}$ ▷ $i_1, ..., i_{t_1} \in \{1, ..., t\}$
7:       $h_i \leftarrow UHF_u(N_i^\rho)$
8:     **end for**
9:     $T = IV \| h_1 \| ... \| h_{t_2}$
10:     **return** T
11: **end procedure**
---

4 A Review of Approximate Message Authentication Codes   119

Algorithm 10 AMAC$_2$ verification [1]

Input: shared secret key k generated by Kg$(.,d)$, initial value IV, m-bit received message x', reference tag T, block size c, approximate correctness parameters p and δ, approximate security parameter γ, threshold value Tr

1: **procedure** AMAC$_2$ − VF(k,x',T)
2:     Write $T = IV||h_1||...||h_{t_2}$
3:     $T' =$ AMAC$_2$ − Tag(k,x')
4:     **if** $h_i = h'_i$ for at least $Tr.t_2$ number of blocks **then**
5:        **return** 1 and stop.
6:     **end if**
7:     **return** 0
8: **end procedure**

The above two mentioned AMACs are the first approximate message authentication schemes where the threshold value or block size can be approximated based on the desired approximate correction and approximate security level theoretically. Also the detailed security analysis and possible attack scenarios have been given in [1]. The main application of the latter mechanisms is biometric entity authentication [1].

4.2.3.1 Analysis of the Computational Security Based AMACs

The above mentioned AMAC approaches have been designed based on the computational power of the attacker and desired performance. The first AMAC (AMAC$_1$) uses a standard MAC in the tag generation and verification algorithms. If an ideal MAC with N-bit output is engaged, it is easy to see that the AMAC$_1$ satisfies $(1, \delta + 1)$-approximate correctness property, $(HD, 1, t - \mathcal{O}(q|pc|), q, 1 - \exp(-\frac{q^2}{2^N}))$-approximate security property and preimage resistance property [1].

The approximate correctness and approximate security properties of AMAC$_2$ have been proven by theorem 2 in [1]. According to that, if the $PRF_k(.)$ is a (t_1, q_1, ϵ_1)-secure pseudo-random function and the UHF is a (t_2, q_2, ϵ_2)-secure target collision resistant hash function, then AMAC$_2$ satisfies (p, δ)-approximate correctness and $(HD, 2\delta, t_3, q_1, \epsilon_1 + \epsilon_2 q_1 + 1 - p)$-approximate security properties in which $t_3 = \min(t'_1, t'_2)$ where t'_1 and t'_2 are defined as follows [1].

$$t'_1 = t_1 - \mathcal{O}(q_1(m \log(m) - \log(1-p)) - \log(1-p)$$
$$t'_2 = t_2 - \mathcal{O}((m \log(m) - \log(1-p)) - \log(1-p)$$

The main interesting property of the AMAC$_2$ is ability to find a lower bound for the threshold based on the requested accuracy and security level. In this regard, we define the following random variables [1, 15].

$$B_{ij} = \begin{cases} 1, & h_i \neq h'_i \\ 0, & h_i = h'_i, \end{cases} \quad (4.15)$$

$$NB = \sum_{i=1}^{t_2} NB_i$$

Using classical occupancy problem [27], the probability of appearing one non-matched block, P_{NB}, AMAC$_2$-Tag is calculated as follows [15].

$$P_{NB} = P(NB_i = 1)$$
$$= 1 - (1 - \frac{\binom{n-t_1+\delta_2-1}{\delta_2}}{\binom{n+\delta_2-1}{\delta_2}})2^{-b} + \frac{\binom{n-t_1+\delta_2-1}{\delta_2}}{\binom{n+\delta_2-1}{\delta_2}} \quad (4.16)$$

To obtain a desirable level of security and accuracy, the probability of appearing less number of non-matched blocks than a threshold value($\lceil Tr.t_2 \rceil$) must be calculated. The corresponding random variable NB follows the Binomial distribution with parameters t_2 and P_{NB}. After applying Chernoff inequality [1, 33] and approximate accuracy requirement, a lower bound for number of required hash blocks [1] or the threshold ratio can be extracted smoothly [1, 15]:

$$Tr \geq P_{NB} + \sqrt{\frac{1}{2t_2} \ln \frac{1}{1-p}} \quad (4.17)$$

4.2.4 Unconditionally Secure AMAC

Safavi-Naini et al., were the first who formalized the theory of approximate authentication and fuzzy universal hashing in unconditionally secure framework [34] based on the earlier work in 2005 [35]. The adversary model has been investigated by finding tight bounds for impersonation and substitution attacks. In this framework no assumption on the computational power of the adversary is made. They generalized the concept of classic universal hashing systems to the fuzzy authentication systems where the partially different messages are indistinguishable from each other. The corresponding security analysis was provided as well. Following this idea [34], Tonion et al. [14] gave the formal definition of unconditionally secure approximate message authentication schemes and their analysis. The proposed abstract construction method based on extracting a random substring from the permuted message was shown to match with unconditional secure settings [14]. This method is known as a formal method for compressing the binary data in fuzzy authentication systems. The mechanism will be called here unconditionally secure AMAC (USAMAC)(Algorithm 11).

Algorithm 11 USAMAC tag generation [14]

Input: shared secret key k generated by $Kg(.,d)$, m-bit message $x = x_1...x_m$
1: **procedure** USAMAC − Tag(k,x)
2: $\quad f_k(x_1...x_m) = (x_{i_1}...x_{i_n})$
3: $\quad T = f_k(\pi_k(x)) = (x_{\pi_k^{-1}(i_1)},...,x_{\pi_k^{-1}(i_n)})$ ▷ $f_k(.)$ is a substring selection function and $\pi_k(.)$ is a permutation function.
4: $\quad$ **return** T
5: **end procedure**

The verification algorithm verifies the authenticity of the message if the received message is close enough to the original message indicated from the threshold value of the distance between two tags. The description of the verification algorithm is given in Algorithm 12.

Algorithm 12 USAMAC tag verification [14]

Input: shared secret key k generated by $\text{Kg}(.,d)$, m-bit message $x' = x'_1...x'_m$, reference tag T, threshold value δ
1: **procedure** $\text{USAMAC} - \text{VF}(k, x', T)$
2:     $T' = \text{USAMAC} - \text{Tag}(k, x')$
3:     **if** $HD(T, T') < \delta$ **then**
4:         **return** 1 and halt.
5:     **end if**
6:     **return** 0
7: **end procedure**

4.2.4.1 Analysis of Unconditionally Secure AMAC

The analysis of USAMC is given in [14] in detail. The security analysis of the scheme has been given from unconditional security viewpoint where no assumptions on the adversary's computational power are given. According to Definition 2, the formal model of USMAC is a (HD, p_m, δ)-approximate correct and (HD, δ, t, 1, ϵ)-approximate secure MAC where the probability p_m is defined as follows [14].

$$p_m = \min_{x \in X, k \in K} Pr\{\text{VF}(k, x', t) = 1, \text{HD}(x, x') < \delta\} \quad (4.18)$$

The bounds for deception probability of the attacker in a passive and an active adversary model have been extracted in [14].

4.2.5 Comparison

The introduced approximate message authentication schemes have been categorized upon their design methods. They are used according to the desired application types which can demand different requirements. Different criteria are concerned for these schemes. The most important of them are robustness, security, error correction, and error localization. The latter one is more desired for application in image authentication but, as mentioned before some schemes may satisfy it as well. The following table summarizes a comparison among the introduced schemes. According to Table 4.1 the N-ary AMAC scheme provides the most robustness among all schemes which deteriorates the security (by increasing the false acceptance rate). Also AMAC_1 [1] and EC-WNTMAC [31] are the only schemes providing error correction capability but the AMAC_1 does not tolerate any error. Finally, the only scheme whose security has been analyzed from unconditional security viewpoint is USAMAC [14].

Table 4.1 Comparison among approximate message authentication categories

Scheme	Robustness	Security	Error correction	Error localization
N-ary AMAC [18]	Highly tolerant	Low	No	No
Repaired N-ary AMAC [16]	Good	Good	No	No
BAMAC [15]	Good	Good	No	No
NTMAC [28]	Good	Computational secure	No	Yes
CRC-NTMAC [29]	Good	Computational secure	No	Yes
EC-WNTMAC [31]	Good	Computational secure	Yes	Yes
$AMAC_1$ [1]	No	Computational secure	Yes	No
$AMAC_2$ [1]	Adjustable	Computational secure	No	No
USAMAC [14]	Adjustable	Unconditionally secure	No	No

4.3 Applications of Dedicated AMACs in Image Authentication Techniques

4.3.1 Extension of the MAJORITY-Based AMACs and the NTMACs

Some of the above introduced AMAC mechanisms have been applied in image authentication methods creatively. The first AMAC mechanism, **2-ary AMAC** which is called image message authentication code (IMAC) in [36] is used to authenticate by dividing the image into non-overlapping rectangular blocks. To generate the tag, the first most significant bits (msb) of the mean of the coefficients of each block is calculated which results in a binary image. Then the **2-ary AMAC-Tag** is applied on the rows and columns which enables partial error localization at the receiver side. However, the erroneous block can not be identified uniquely. An error tolerance zone is estimated to introduce more robustness into the image authentication mechanism. Let Tr be the maximum allowed absolute difference between the original and modified pixel value of a gray scale image. The interval [0, 255] is split into two subintervals $I_1 = [0, 127 - Tr]$ and $I_2 = [128 + Tr, 255]$ and then all pixel values in intervals [0, 127] and [128, 255] are linearly mapped into I_1 and I_2 respectively. The created symmetric gap of length $2Tr + 1$ results in more robustness in average values of the pixels. The original image firstly is subjected to this linear mapping prior submitting.

IMAC tolerates the JPEG compression and additive Gaussian noise image modifications [36]. So, the robustness against non-malicious image modifications like partial rotation or scaling is very limited. Also the security of the scheme against image tampering is not strong as only the limited number of pixel bits are used to generate the image feature. However, the authors suggested to use more msbs to enhance the security [36].

Like the first AMAC, the NTMAC-family AMACs have been also used in image authentication methods directly. In [37], the **NTMAC** is applied on the image after different image partitioning. **NTMAC-Tag** calculates the tag of the image on the deterministically or randomly partitioned image. The introduced scheme is called image NTMAC or simply INTMAC. The advantage of different partitions is increasing the error localization capability. Very similar approach has been proposed in [38] in which **CRC-NTMAC-Tag** is used instead.

4.3.2 Extension of $AMAC_1$ and $AMAC_2$

In a very recent approach, the schemes defined based on the computational security approach in 4.2.3.1 have been used in a combination form for the sake of image authentication [15]. The scheme is based on the DCT coefficients of the image blocks and is called approximate authentication and correction of images (AACI). Unlike the IMAC, the tag generation algorithm of the AACI is identical for both sender and receiver [15]. The image I which is supposed to be transmitted over a non-secure channel is divided into $M \times N$ disjoint rectangular blocks each of $m \times m$ pixels size (here $m = 8$). The DCT is applied to each pixel block and the DC element of each matrix is selected. The selected DCs result in an lMN-bit sequence where l is the length of each represented quantized DC element. The $AMAC_1$-Tag is used for the tag generation corresponding to DCs bit-sequence.

The tag generated by $AMAC_1$-Tag is denoted by $T1$. To generate a final tag, the first ten AC coefficients of the DCT matrix of each block are selected according to zigzag order used in the JPEG compression. The resulted $10l$-bit sequence is considered as a message block (out of MN blocks) according to $AMAC_2$-Tag which is used to generate the authenticate tag of the $10lMN$-bit message which results in subtag $T2$. The final tag is concatenation of the two subtags, i.e., $T = T1\|T2$.

In this scheme, the corresponding adjustable parameters are set according to desired accuracy and security level and image application sensitivity. Protecting the first ten AC coefficients along with the DC coefficient which have sufficient information of an image block ensures acceptable security against common types of image tampering attacks [15, 39].

The verification process of the AACI is as follows [15]. The receiver gets the compressed image with the attached AACI tag. The image is decompressed and the AACI tag is computed on the image based on the shared secret keys and compared to the received one. If the verification succeeds on the first subtag $T1$, then the verification on the second part $T2$ starts. Otherwise, the detected errors are tried to be corrected by the error-correcting code. The verification procedure on $T2$ is successful when at least $Tr.t_2$ positions have equal hash blocks.

The tag generation algorithm and the verification algorithm are presented in Algorithm 13 and Algorithm 14 respectively.

Algorithm 13 AACI tag generation [15]

Input: image I, image dimensions M, N, error correction capability of $AMAC_1$ δ_1, shared secret keys k_1, k_2, initial value IV, approximate correctness parameters p and δ_2, approximate security parameter γ

1: Let $\{b_{1,1},...,b_{M,N}\}$ be image blocks of I.
2: **procedure** AACI $-$ Tag(k_1,k_2,I)
3:     **for** $i = 1, M$ **do**
4:         **for** $j = 1, N$ **do**
5:             A=DCT$(b_{i,j})$ ▷ DCT applies to each image block.
6:             $DC_{i,j} \leftarrow A_{1,1}$ ▷ DC of each image block is extracted.
7:             $AC_{i,j} \leftarrow A_{1,2}||A_{2,1}||...||A_{5,1}$ ▷ ACs of each image block are extracted.
8:         **end for**
9:     **end for**
10:    Let $M_1 = \{DC_{1,1},...,DC_{M,N}\}$
11:    Let $M_2 = \{AC_{1,1},...,AC_{M,N}\}$
12:    $T1 = AMAC_1 - $ Tag(k_1,M_1)
13:    $T2 = AMAC_2 - $ Tag(k_2,M_2)
14:    $T = T1||T2$
15:    **return** T
16: **end procedure**

The security and performance of the AACI are discussed in [15] and Subsection 4.2.3.1 in detail. However, an existing potential threat in the hybrid scheme AACI is the impact of image tampering. The verification could be successful on a valid tampered image which is not authentic if the tampered pixels are not reflected by the protected DCT components. Protecting 11 DCT matrix elements of each image block makes this attack scenario very hard. However, this is not proven for the AACI and comprehensive analysis as the impact of image tampering is required [15].

Another security flaw is that the a collision between the MAC blocks may happen as a result of error correction. This problem can be avoided if the MAC by $AMAC_1$ is calculated on the DCs while, some least significant bits (lsb) are protected by the error-correcting code instead of the whole DC part as in the original scheme.

Algorithm 14 AACI verification [15]

Input: decompressed received image I', image dimensions M, N, reference tag T, shared secret keys $k_1 = k_{1,a} \| k_{1,e}$, k_2, initial value IV, δ_1, approximate correctness parameters p and δ_2, threshold ratio Tr, approximate security parameter γ

1:   $t_1 = \lceil MN/2\delta_2 \rceil$, $t_2 = \lceil -10\log(1-p) \rceil$
2:   Let $\{b'_{1,1}, ..., b'_{M,N}\}$ be image blocks of I'.
3:   **procedure** AACI $-$ VF(k_1, k_2, I')
4:     **for** $i = 1, M$ **do**
5:       **for** $j = 1, N$ **do**
6:         A=DCT$(b'_{i,j})$
7:         $DC'_{i,j} \leftarrow A_{1,1}$
8:         $AC'_{i,j} \leftarrow A_{1,2} \| A_{2,1} \| ... \| A_{5,1}$
9:       **end for**
10:     **end for**
11:     Let $M'_1 = \{DC'_{1,1}, ..., DC'_{M,N}\}$
12:     Let $M'_2 = \{AC'_{1,1}, ..., AC'_{M,N}\}$
13:     Write $T = T1 \| T2$.
14:     Let M'' be the extracted message from T_1 using $k_{1,e}$.
15:     **if** $MAC(k_{1,a}, M'_1) = MAC(k_{1,a}, M''_1)$ **then**
16:       **go to** 24
17:     **else**
18:       **if** Error correction on M''_1 is possible **then**
19:         Perform error correction and **go to** 15
20:       **else**
21:         **return** 0 and stop
22:       **end if**
23:     **end if**
24:     $T'2 = $ AMAC$_2 -$ Tag(k_2, I')
25:     Write $T2' = IV \| h'_1 \| ... \| h'_{t_2}$ and $T2 = IV \| h_1 \| ... \| h_{t_2}$
26:     **if** $h_i = h'_i$ for at least $Tr.t_2$ number of blocks **then**
27:       **return** 1
28:     **end if**
29:     **return** 0
30: **end procedure**

4.4 Conclusion

In this chapter, a concrete overview of the generic approximate message authentication codes has been given. They have been categorized into different groups according to their design methods. The security analysis and robustness study is dedicated for each group so different performance is achieved. The theory of approximate message authentication codes which lies in the field of robust message authentication is a new theory and much time is needed to achieve its maturity. The importance of these AMACs is emphasized mainly because of its broad range application fields. Multimedia objects or biometric data have a fuzzy nature and their authentication method should be focused on their contents rather than the whole representation. Otherwise, a reasonable performance can not be gained.

There are too many approaches to AMAC designs introduced so far. However, it is not possible yet to analyze their security in a systematic way like the known cryptographic primitives. The extension of the security analysis from cryptography to this field is of a great importance as it fills a big gap in their theory. Future efforts will be concentrated toward this goal.

References

1. Crescenzo GD, Graveman RF, Arce GR. Approximate message authentication and biometric entity authentication. In: Proc. Financial Cryptography, LNCS. 2005; vol. 3570, pp. 240–54.
2. Xie L, Arce GR, Graveman RF. Approximate image message authentication codes. IEEE Trans Multimed. 2001;3(2):242–52.
3. Lin CY, Chang SF. SARI: self authentication and recovery image watermarking system. ACM Multimed. 2001;4518:628–9.
4. Swaminathan A, Mao Y, Wu M. Robust and secure image hashing. IEEE Trans Inf Forensics Secur. 2006;1(2):215–30.
5. Lin CY, Chang SF. A robust image authentication method surviving JPEG lossy compression. In: Proc. SPIE Storage and Retrieval of Image/Video Database, San Jose; 1998.
6. Haouzia A, Noumeir R. Methods for image authentication: a survey. Multimed Tools Appl. 2008;39:1–46.
7. Ur-Rehman O, Zivic N. Noise tolerant image authentication with error localization and correction. In: Proc. 50th Annual Allerton Conference on Communication, Control and Computing, Illinois, USA; 2012.
8. Ur-Rehman O, Tabatabaei SAE, Zivic N, Ruland C. Soft authentication and correction of images. In: Proc. 9th International ITG Conference on Systems, Communications and Coding (SCC 2013), Munich, Germany; 2013.
9. Jing F, Li M, Zhang HJ, Zhang B. An efficient and effective region-based image retrieval framework. IEEE Trans Image Process. 13(5):699–709; 2004.
10. Queluz MP. Toward robust, content based techniques for image authentication. In: Proc. Second Workshop on Multimedia Signal Processing. 1998; pp. 297–302.
11. Kailasanathan C, Safavi-Naini R, Ogunbona P. Image authentication surviving acceptable modifications. In: IEEE-EURASIP, Workshop on Nonlinear Signal and Image Processing; 2001.
12. Chang IC, Hsu BW, Laih CS. A DCT quantization-based image authentication system for digital forensics. In: Proc. First Int. Workshop on Systematic Approaches to Digital Forensic Engineering. 2005; pp. 223–5.
13. Chang HT, Hsu C-C, Yeh C-H, Shen D-F. Image authentication with tampering localization based on watermark embedding in wavelet domain. Optical Eng. 2009;48(5):057002.
14. Tonien D, Safavi-Naini R, Nickolas P, Desmedt Y. Unconditional secure approximate message authentication. In: Proc. 2nd International Workshop on Coding and Cryptology, LNCS. 2009; vol. 5557, pp. 233–47.
15. Tabatabaei SAE, Ur-Rehman O, Zivic N. AACI: a mechanism for approximate authentication and correction of images. In: Proc. International Conference on Communication (ICC 2013), Budapest, Hungary. 2013; pp. 727–32.
16. Tonien D, Safavi-Naini R, Nickolas P. Breaking and repairing an approximate message authentication scheme. Discret Math Algorithms Appl. 2011;3(3):393–412.
17. Graveman RF, Fu K. Approximate message authentication codes. In: Proc. 3rd Annual Symposium on Advanced Telecommunication and information Distribution Research Program (ATIRP), USA; 1999.
18. Ge R, Arce GR, Crescenzo GD. Approximate message authentication codes for N-ary alphabets. IEEE Trans Inf Forensics Secur. 2006;1(1):56–67.
19. De A, Mossel E, Neeman J. Majority is stablest. http://arxiv.org/pdf/1211.1001v2.pdf. Accessed 2 Nov 2012.

20. Tabatabaei SAE, Zivic N. Revisiting a primitive: analysis of approximate message authentication codes. In: Proc. International Conference on Communication (ICC 2014), Sydney, Australia. 2014; pp. 743–748.
21. Daum M. Cryptanalysis of Hash functions of the MD4-family. Ph.D. thesis. Ruhr-Universität Bochum, persistent identifier: urn:nbn:de:hbz:294-14245; 2005.
22. Makri FS, Psillakis ZM. On success runs of a fixed length in Bernoulli sequences: exact and asymptotic results. Comput Math Appl. 2011;61:761–72.
23. Makri FS, Psillakis ZM, Kollas N. Counting runs of ones and ones in runs of ones in binary strings. Open J Appl Sci. 2012;2(4B):44–7.
24. Benson G. Tandem repeat finder: a program to analyze DNA sequence. Nucl Acid Res. 1999;27:573–80.
25. Nuel G, Regad L, Martin J, Camprous AC. Exact distribution of a pattern in a set of random sequences generated by a Markov source: applications to biological data. Algorithms Mol Biol. 2010;5:1–18.
26. http://csrc.nist.gov/publications/nistpubs/800-22-rev1a/SP800-22rev1a.pdf.
27. Feller N. An Introduction to probability theory and its applications. 3rd ed. New York: Wiley; 1968.
28. Boncelet CG Jr. The NTMAC for authentication of noisy messages. IEEE Trans Inf Forensics Secur. 2006;1(1):35–42.
29. Liu Y, Boncelet CG Jr. The CRC-NTMAC for noisy message authentication. IEEE Trans Inf Forensics Secur. 2006;1(4):517–23.
30. Liu Y, Boncelet CG Jr. The BCH-NTMAC for noisy message authentication. In: Proc. 40th Annual Conference on Information Sciences and Systems; Mar. 2006, pp. 246–51.
31. Ur-Rehman O, Zivic N, Tabatabaei AE, Ruland C. Error correcting and weighted noise tolerant message authentication codes. In: Proc. 5th International Conference on Signal Processing and Communication Systems (ICSPCS), USA; 2011.
32. Naor M, Yung M. Universal one-way hash functions and their cryptographic applications. In: Proc. ACM-STOC; 1989.
33. Mitzenmacher M, Upfal U. Probability and computing—randomized algorithms and probabilistic analysis. USA: Cambridge University Press; 2005.
34. Safavi-Naini R, Tonien D. Fuzzy universal hashing and approximate authentication. Discret Math Algorithms Appl. 2011;3(4):587–607.
35. http://eprint.iacr.org/2005/256.
36. Arce GR, Graveman RF. Approximate image message authentication codes. IEEE Trans Multimed. 2001;3(2):242–52.
37. Boncelet C. Image authentication and tamper proofing for noisy channels. In: Proc. ICIP. 2005; vol. 1, pp. 677–80.
38. Liu Y, Boncelet C. The CRC-NTMAC for image tamper proofing and authentication. Proc. IEEE international Conference on image Processing. 2006; vol. 1, pp. 1985–8.
39. Zhu BB, Swanson MD, Tewfik AH. When seeing is not believing. IEEE Signal Mag. 2004;21(2):40–9.

Chapter 5
Fuzzy Image Authentication with Error Localization and Correction

Obaid Ur-Rehman and Nataša Živić

5.1 Introduction

During the transmission of multimedia data from source to sink, multiple elements may contribute to data alteration. These elements typically include quantization mechanisms inside the source encoder (or decoder), compression mechanisms, such as lossy compression, and the channel noise, such as the one induced by a wireless medium. In order to cope with such forms of data modifications, communication systems typically employ error correcting codes, which add protection (parity) bits to the original data in such a manner that some or all of the modifications (or errors) in the data can be corrected. Some of the most widely used error correcting codes include Reed–Solomon (RS) codes [1], turbo codes [2] and low density parity check (LDPC) codes [3]. These error correcting codes are good at correcting a certain predefined number of bit modifications. However, if the number of errors exceeds the error correction capability of the codes, they are not correctable. In such a situation, different mechanisms are employed by the data transmission protocols for data recovery. Automatic Repeat reQuest (ARQ) [4] (or its variant called Hybrid-ARQ) is an example of such protocols, where the erroneous data is retransmitted up to a few times, until either the data is received as intended or a threshold number of retransmissions have been done. Most of these communication protocols use error detection codes, such as cyclic redundancy code (CRC) [5] or a message authentication code (MAC) [6, 7], to verify the integrity of data. Hashed message authentication code (HMAC) [8], such as SHA-2, is used as a means to authenticate the origin of the data in addition to its integrity.

Standard MAC and HMAC algorithms are designed such that even a single bit modification in the data changes its MAC by almost 50 %. These authentication algorithms are "hard authentication" algorithms, where the authentication decision is a strict "NO" when the data or its MAC has changed even by a single bit. In such

---

O. Ur-Rehman (✉) · N. Živić
Chair for Data Communications Systems, University of Siegen, 57076 Siegen, Germany
e-mail: obaid.ur-rehman@uni-siegen.de; natasa.zivic@uni-siegen.de

© Springer International Publishing Switzerland 2015
N. Živić (ed.), *Robust Image Authentication in the Presence of Noise*,
DOI 10.1007/978-3-319-13156-6_5

a case, the transmitted image (or a videoframe) will not be accepted as authentic at the receiver. In case of multimedia, such as an image or a videoframe, a few bit modifications might not have any visual impact on the content of the multimedia due to the nature of multimedia and the human visual perceptual system. For example, a few modifications in the least significant bits of an image data might not change the content and a bare human eye might not be able to differentiate between the original and modified image. If it is not possible to perform retransmissions using (H-)ARQ, e.g., either there is no feedback channel or a real-time communication is intended, or if the number of errors exceeds the error correction capability of the channel codes, the erroneous multimedia content will usually be discarded if hard authentication algorithms are used. In such a case, the receiver might interpolate or extrapolate the data, e.g., in a video stream the previous frame might be re-played or a blank frame might be inserted between the previous and the next frame. This is due to the fact that in some applications, it is be better to have partial data than no data at all.

In order to solve this issue, fuzzy authentication algorithms [9] have been introduced in literature. Primarily, these algorithms include approximate message authentication code (AMAC) [10], image message authentication code (IMAC) [11] and noise tolerant message authentication code (NTMAC) [12]. Some fuzzy as well as standard hard authentication algorithms, tailored specifically for multimedia data, have also been proposed [13–16]. In most of the cases, these algorithms work in little or no coordination with the other modules of the communication systems. This means normally the source coding module has little to do with the cryptographic module and the cryptographic module performs in little coordination with the channel coding. Recently, it has been shown that a better coordination between the different components of communication system improves the channel coding results as well as the authentication results [17]. Some recent fuzzy authentication algorithms, where different components of a communication system interact for authentication as well as error localization and recovery from some errors, are weighted noise tolerant authentication code (WNTMAC) [15], error correcting noise tolerant message authentication code (EC-NTMAC) [18, 19]. In WNTMAC, a different weight is assigned to important and non-important parts of a message. Important parts are provided more protection than the non-important parts. More efforts are spent on the important blocks to identify and correct errors, whereas lesser efforts are spent on the data parts of lesser importance. In EC-NTMAC, the NTMAC algorithm for fuzzy authentication is extended and error correction capability is integrated into the fuzzy authentication algorithm to do error correction in addition to fuzzy authentication. Applications in image authentication together with simulation results are given in [18, 19].

This chapter is organized as follows. In Sect. 5.2, a brief overview of the building blocks used in the fuzzy image authentication algorithms is given. This includes the frequency domain transformation, such as discrete cosine transform and error correcting codes such as turbo codes. In Sect. 5.3, applications of error correcting codes in image authentication are discussed. In Sect. 5.4, two fuzzy authentication algorithms for image authentication based on the integration of error correcting codes are discussed. The performance and security analysis of these algorithms is given in Sect. 5.5. Finally, simulation results are shown in Sect. 5.6 to show the performance.

5.2 Building Blocks of the Fuzzy Image Authentication Algorithms

5.2.1 Content Based Authentication

In order to have the capability to authenticate multimedia data, such as images or videoframes (this chapter discusses only image authentication from this point onward), despite certain changes in the original data, it is advantageous to base the authentication on the content rather than the actual data. The actual data of an image, such as the bit or byte values might change due to quantization, compression, etc. The features of an image are, however, not changed by small modifications. Image authentication based on its content or distinct features, rather than the actual data, is called content-based authentication. Different feature extraction techniques have been proposed in literature by the image processing community. The features can be as simple as the image contour, edges, Fourier, wavelet, or cosine transform coefficients or more complex features such as those based on the image textures. The fuzzy authentication algorithms discussed in this chapter are based on the frequency domain features obtained using the discrete cosine transform (DCT) [22] of an image.

5.2.2 Discrete Cosine Transform

DCT [22] is one of the most widely used techniques in image processing. Other widely used transforms are the discrete Fourier transform (DFT) and the discrete Wavelet transform (DWT). Chen and Pratt [20] pioneered the application of DCT in image processing. DCT removes correlation from image data, after which each transform coefficient can be encoded independently without devitalizing compression efficiency [21]. Since DCT combines most of the energy in a few transform elements, it is preferred as compared to DFT.

DCT of a matrix (called a block in image processing terminology) is defined as follows [22, 23]:

$$X(l,k) = \alpha(l)\alpha(k)\frac{2}{N}\sum_{m=0}^{N-1}\sum_{n=0}^{N-1} x(m,n)\cos\left[\frac{\pi(2m+1)}{2N}\right]\cos\left[\frac{\pi(2n+1)}{2N}\right]. \quad (5.1)$$

where,

$$\alpha(i) = \begin{cases} 1/\sqrt{2}, & i = 0 \\ 1, & 1 \leq i \leq N-1 \end{cases}$$

It can be observed from (5.1) that the element at the index (0, 0) of the DCT (called DC element) represents the average intensity of the corresponding block and contains most of the energy and perceptual information. Elements at the indexes other than

(0, 0) are known as AC elements. AC elements in the upper minor diagonals have more information as compared to the AC elements in the lower minor diagonals. If the DC components together with a reasonable number of AC components are retained, e.g., by passing the DCT elements through a low-pass filter, the inverse transform of the DCT (IDCT) can reconstruct the original block with some or no loss in quality. By increasing the number of AC components, the loss in the quality of reconstruction decreases.

The inverse of a 2-D DCT for l, k = 0, 1, ..., N − 1 can be calculated as follows,

$$x(l,k) = \alpha(l)\alpha(k)\frac{2}{N}\sum_{m=0}^{N-1}\sum_{n=0}^{N-1} X(m,n)\cos\left[\frac{\pi(2m+1)}{2N}\right]\cos\left[\frac{\pi(2n+1)}{2N}\right] \quad (5.2)$$

5.2.3 Error Correcting Codes

5.2.3.1 Error Correcting Codes in Image Authentication

Error correcting codes, such as RS codes, turbo codes, or LDPC codes, are used to detect and correct the errors introduced in the data by different means, such as by the transmission noise over a communication channel or during storage over a storage medium. The choice of error correcting codes depends on the noise environment and the desired error correction capability. RS codes are good at correcting burst errors in addition to the sporadic errors and work more naturally on symbols. LDPC codes and turbo codes operate on large data sets such as multimedia data and their error correction capability has been demonstrated to be very closer to the Shannon limit [24]. Error correcting codes, when combined with image authentication, increase the tolerance of the authentication algorithm. They also help in identification of the error location and correcting the errors if they are within the error correction capability. Mostly such errors are due to unintentional modifications and therefore in a small number. In this section, a short introduction of some of most widely used error correcting codes is given. These codes are also used in the algorithms discussed next and therefore, they are introduced here before their application is discussed.

5.2.3.2 Reed–Solomon Codes

RS codes were proposed by Reed and Solomon in their legendry paper [1]. RS codes are nonbinary cyclic codes. RS codes work on symbols rather than bits, where each symbols is a Galois Field (GF) element. Each symbol is made up of m-bits, such that m is an integer and usually $2 \leq m \leq 8$. The m-bit elements are defined over the Galois Field GF(q), where $q = p^m$ and p is a prime number typically chosen to be 2. RS codes are linear block codes, which mean that they encode data in blocks. An RS(n, k) code takes a data block of k symbols and protect it by appending n − k parity symbols to produce a block of n symbols.

Here,

$$0 < k < n < p^m + 2 \tag{5.3}$$

The error correcting capability of RS codes is measured in symbols instead of bits. Thus, a burst of bit errors can be grouped in to fewer symbols and then corrected as symbols. The choice of the number of parity symbols dictates how many errors should be corrected by the RS(n, k) codes. If t symbol errors are desired to be corrected by the RS(n, k) code, then n − k must be chosen such that $2t \leq n - k$.

RS codes are also good at correcting erasures, where erasures are the errors whose positions are known in advance via some a priori knowledge. The joint error and erasure correction capability of RS codes is expressed as,

$$2N_e + N_r \leq n - k \tag{5.4}$$

where N_e is the number of errors and N_r is the number of erasures. (5.4) means that if t symbol errors are to be corrected (i.e., $N_e = t$ and $N_r = 0$), then 2t parity symbols are needed. If only erasure correction is desired, then erasures equal to twice the number of errors are correctable, i.e., up to 2t erasures can be corrected.

The decoding algorithm for RS codes can either be a hard or a soft decision decoder. Hard decision decoders are the ones that operate on discrete values. Most widely used hard decision decoding algorithms for RS codes are Berlekamp–Massey [25] algorithm and the Euclidean algorithm [26]. Soft decision decoders, on the other hand, operate on continuous values or floating point numbers. They usually take soft values at the input and generate soft values at the output (soft input soft output (SISO) decoders). However, in general they can also generate only hard output. Well-known soft decision decoding algorithms for RS codes are Guruswami–Sudan [27] decoding algorithm and the Koetter-Vardy [28] decoding algorithm. A more generalized approach is to output a list of codewords when it is not possible to decide on one codeword, e.g., multiple codewords are equiprobable to have been transmitted, given the received word. This approach is called list decoding [29].

Since the hard decision decoders operate on the quantized data, where some information is lost during quantization, their decoding performance is typically not as good as that of the corresponding soft decision decoder [9].

5.2.3.3 Turbo Codes

Turbo codes, proposed by Berrou, Glavieux, and Thitimajshima in 1993, were the first practical codes shown to approach the channel capacity [2]. Turbo codes are theoretically a concatenation of two codes connected by an interleaver. In practice, the constituent codes are usually convolutional codes connected in parallel. For encoding, the input data is interleaved and passed through the encoders. Without loss of generality, it can be said that the input data is passed through an identity interleaver and then through the first encoder and in parallel the input data is passed through another (nonidentity) interleaver before the second encoder. The result is the original data and two encoded sequences or the parity bits.

The decoder for turbo codes is usually a soft decision iterative decoder. The decoder's task is to find the most probable codeword based on the demodulated received sequences and a priori knowledge or probabilities about the messages and their occurrence. The decoding in turbo codes is performed by two decoders, each one capable of working on soft input and producing soft output—or the likelihood information about the output bits. The most common SISO decoding algorithms are soft output viterbi algorithm (SOVA) and the maximum a posteriori probability (MAP) decoding algorithm. The decoding is performed in turns by the constituent decoders. It starts with one decoder and the output of the decoder is an estimate of the first codeword. This estimate is used as a priori information in improving the decoding results of the second decoder, which works on the second encoded sequence and the interleaved output of the first decoder. The used interleaver is the same as the one used in the encoding. This decoding process is repeated iteratively, each time improving the confidence level in the decoded sequence and increasing the reliability of the decoded bits. Repeating the iterations many times would result in a convergence to a final decision about the most probable codeword.

The likelihood of the input and the output bits in the decoders are normally represented as log likelihood ratios (LLRs). The LLR of each bit is based on the a priori probability of the bit and the current observation of the bit. If the sequence to be decoded is $x = (x_1, x_2, x_3, \ldots, x_n)$, then the LLR of each bit x_i can be represented as,

$$LLR(x_i) = \log \frac{\Pr(x_i = 1, observation)}{\Pr(x_i = 0, observation)} \quad (5.5)$$

It can be noticed from (5.5) that the LLR of a bit is based on the a priori probabilities of that bit and the current observation of the bit value.

5.3 Applications of Error Correcting Codes in Image Authentication

In practice, it is important to not only authenticate an image and verify its integrity but also to locate any modifications in the image. Existence of modifications can be detected using cyclic redundancy check (CRC) [5], MACs [6, 7], digital signatures or digital watermarking [30]. Additionally, it is also very important to locate the exact positions of modifications, to as fine grained level as possible, to find out which objects in the image have changed. However, recently it is getting important to look for methods to do error correction in addition to error detection and localization. Error correcting codes, such as RS and turbo codes, have been used in image authentication algorithms to get an additional error localization and correction capability. They have been widely used in detecting errors in the image, isolating the (potentially) erroneous locations and if possible correcting those (potentially) erroneous parts of the image. When combined with the authentication mechanisms, the error correcting codes can help in partial image recovery even if complete image recovery is not possible. This is particularly helpful in multimedia streams, where a partially authentic image (or part wise authentic image) might be more useful than having no image at all.

In [11], approximate image message authentication codes (A-IMACs) approach for fuzzy image authentication is proposed. The proposed A-IMAC is tolerant to small image modifications but at the same time it is capable of detecting and locating intentional tampering. A-IMAC is based on different composite techniques such as block smoothing, block averaging, parallel AMAC and image histogram enhancement. The performance of A-IMAC in discussed in [11] for three image modification scenarios, i.e., JPEG compression, image forgery, and additive Gaussian noise.

Digital watermarking methods for image authentication with error detection and reconstruction, based on error correcting codes, have also been proposed in literature [31–33]. In [31], RS codes are used to generate parity symbols for each row and column of an image. The resultant parity symbols (as bits) are then embedded as a watermark in the two least significant bit planes of the image. If the watermarked image changes, such that the RS decoder can correct the changes, they are corrected to restore the original image data. A scrambling method is used such that a burst of data modifications is transformed into random noise.

In [32], a watermarking approach is proposed in which the image is divided into blocks and the block hash is encoded with a systematic error correcting code. The parity symbols are then embedded into the blocks. During verification, the hash of each block is recovered with the embedded parity symbols, if the number of tampered blocks does not exceed a threshold value. The size of parity symbols is smaller than the total size of block hashes.

In [33], two techniques for self-embedding an image in itself are introduced with the aim to protect the image content. The self-embedding helps in recovering portions of the image which are cropped, replaced, damaged, or tampered in general. In the first method, the 8×8 blocks of an image are transformed into frequency domain using DCT and after quantizing the coefficients they are embedded in the least significant bits of other, distant blocks. This method provides a good quality of reconstruction but it is very fragile and can be classified as fragile watermarking technique [30]. The second method is based on the principle similar to differential encoding where a circular shift of the original image with decreased color depth is embedded into the original image. It is shown that the quality of the reconstructed image degrades with the increasing level of noise in the tampered image. This method can be classified as a semi-robust watermark [30].

5.4 Fuzzy Image Authentication Codes with Error Localization and Correction

5.4.1 Fuzzy Authentication Based on Image Features

The algorithms introduced in this section (for error correction and soft authentication) are based on image features extracted in the transform domain. The aim is to authenticate the images in the presence of minor modifications and to be able to locate and correct those modifications before declaring the image authentic or unauthentic. Another aim is to be able to locate major modifications in the images protected with the proposed fuzzy authentication algorithms. The features extracted from images are protected by error correcting codes so that minor changes in the features can be reconstructed in case of noise or distortions. This means that as long as the features are correct or correctable, the image is considered authentic, but if the features are distorted to an extent that the reconstruction is not possible, then the image is declared unauthentic. There are many other ways of capturing the features of an image, such as those based on edge detection to capture the contours of image objects, corner detection, blob detection, affine invariant features, image gradient-based features, etc. [34].

The features based on the frequency domain transform, such as DFT, DWT, or DCT are usually extracted by splitting the image into equal sized smaller blocks. Typical dimensions of a block used in practice are 8 × 8 pixels, such as in JPEG and JPEG2000 [35, 36]. Then the desired transform is applied on image block by block. DCT is used in this chapter for frequency domain transformation. The high energy components in the DCT of a block capture the most essential features. Thus, if a "reasonable number" of high energy components are retained, the block can be reconstructed through the inverse DCT (IDCT). This reasonable number determines the quality of reconstruction. If the number of retained components is too small, the quality of reconstruction is not satisfactory. If the number of retained components is increased, the quality of reconstruction is improved. This number also determines the quality of compression and decompression and is represented through the different quantization matrices in JPEG2000.

5.4.2 Image Error Correcting Column-Wise Message Authentication Code (IECC-MAC)

The algorithms for error correction and fuzzy authentication of images introduced in this chapter are based on DCT. The first algorithm is called IECC-MAC [18]. It protects the DC coefficients (as they are the most important) of an image by standard MACs and performs fuzzy authentication with error tolerance on the received (noisy) images. The algorithm is able to localize errors to a smaller part of the modified image and to correct a certain number of these modifications. The algorithms for sender and the receiver side are explained below. Definition 1 is essential to understand the concept of the proposed algorithms,

Definition 1 Suppose that a message M is transmitted over a noisy medium along with its n-bit MAC, denoted as H. Let M' be the received message and H' be the received MAC. Let H'' be the MAC recalculated on M' and let d be a small positive integer (d « n). M' is called d-fuzzy-authenticated if HD(H', H'') ≤ d, where HD is the Hamming distance.

This method of fuzzy authentication, where the Hamming distance between H′ and H″ does not have to be equal to zero for message acceptance, is based on the fact that the MACs of two different messages cannot be the same or even close to each other, in order to avoid near collision attacks [16, 37].

For the sake of simplicity, let's assume that an N × N pixel image has to be transmitted. For other dimensions, padding or row/column repetitions etc. can be used. Let the image be divided into m × m pixel blocks such that m|N, where both N and m are positive integers and m is typically equal to 8. DCT is calculated for each block and the DC components are chosen for protection, by storing them in a DC matrix corresponding to the DC elements of the whole image. This process is described in Fig. 5.1.

Algorithm: IECC-MAC Tag Generation

Input:

- Image I
- Image width W
- Image height H
- Block length m

Algorithm:

blocks = SplitInBlocks(I, W, H, m)

for i = 1 to W/m
    dcs_column$_i$ = [ ]
    for j = 1 to H/m
        dct = ComputeDCTOnBlock(blocks$_{i,j}$)
        dcs_column$_i$ = dcs_column$_i$ || dct$_{1,1}$
    end
    C-MAC$_i$ = ComputeMAC(k$_1$, dcs_column$_i$)
end

IECC-MAC = C-MAC$_1$ || C-MAC$_2$ || .... || C-MAC$_{W/m}$

Output:

    NTMAC$_{DC}$

Standard MAC is computed column-wise on the DC matrix (one MAC for each column of DC elements) to get a total of N/m MACs. These MACs are combined together to get a column-wise MAC (C-MAC) and transmitted together with the compressed image. A compression is achieved by keeping the DC coefficient and the first few minor diagonals of each DCT matrix, which can be chosen according

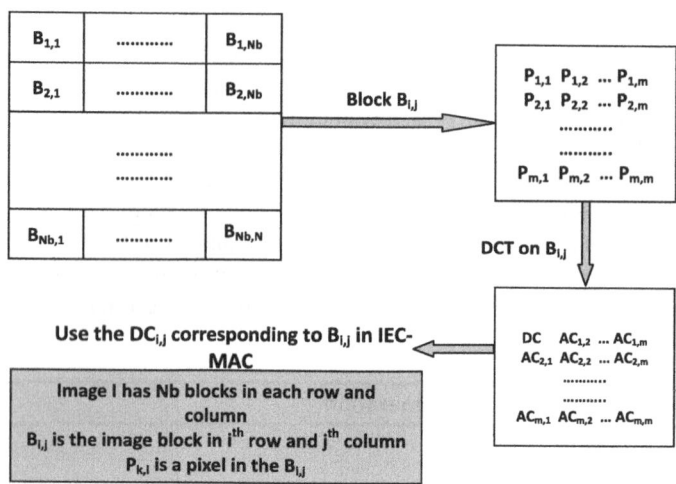

Fig. 5.1 Perform DCT block-wise and keep the DC element.

to the desired image quality as is done in JPEG [35]. An inverse DCT (IDCT) of each DCT matrix (with the DC element and reduced number of AC elements) gives a compressed image. The process of calculation of C-MACs and then combining them to get IECC-MAC is shown in Fig. 5.2.

The receiver receives the (potentially noisy) compressed image I′ and its (potentially noisy) C-MAC′. The receiver recalculates a C-MAC (denoted as C-MAC″) on the image I′ using the same steps as above. C-MAC′ is then compared with C-MAC″ for each of the component MACs. If the Hamming distance between the recalculated MAC (denoted as H″) and the received MAC (called H′) of a column of DC elements is lower than d, i.e., if M′ is not d-fuzzy-authenticated according to definition 1, the corresponding column of blocks in the image I′ is marked as suspicious, otherwise the column is declared to be authentic. The marked columns are called suspicious as they might potentially have errors. After the suspicious columns are marked, they are tried to be corrected using the bit reliability values or the LLRs of each bit.

The LLRs can be obtained from the SISO channel decoder, for example the MAP decoder for turbo codes. If no channel decoder is used, then the channel measurements are taken as the LLRs of each bit. Based on these LLRs and the received signal values, error correction is attempted. The process of error correction works as follows: The bits of the marked suspicious columns are sorted based on their absolute values of LLRs. A combination of least reliable bits is then flipped, followed by recalculation of the MAC (H″) on the DC components of the (bit-flipped) column and calculating its Hamming distance with H′. If the bit-flipped form of marked column is d-fuzzy-authenticated, the corresponding column of blocks in the received image (I′) is accepted as authentic; otherwise, the iterations are continued till a maximum (threshold) number of iterations have been performed. The error localization and correction is shown in Fig. 5.3.

Fig. 5.2 IECC-MAC tag calculation on the DC elements of columns of blocks.

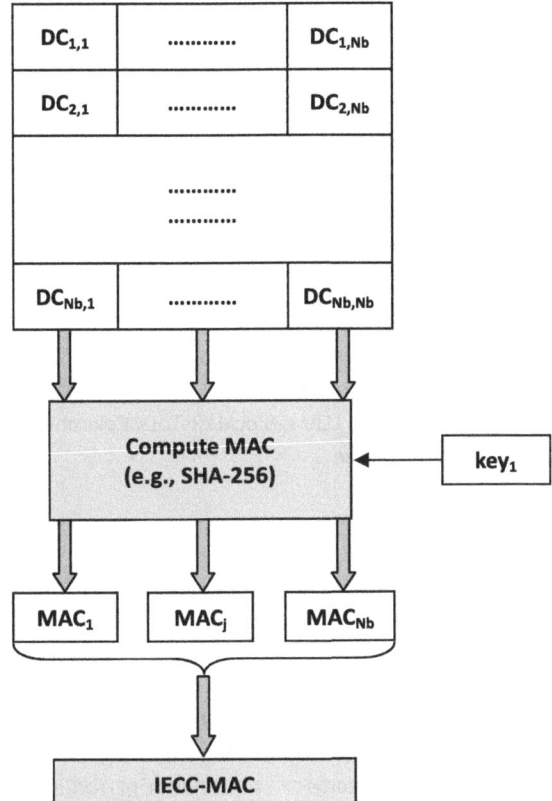

The threshold on the maximum number of iterations (T_{itr}) is predefined and is given by,

$$T_{itr} = 2^{\eta} \qquad (5.6)$$

where η is a small positive integer (typically, $0 \leq \eta \leq 24$). T_{itr} can be made adaptive, by choosing the value of η depending on the bit error rate (BER) and also based on the application demand.

The error localization property of IEC-MAC algorithm needs to be further optimized because IEC-MAC localizes errors to the column level. It would be better to localize errors, correct them, and perform fuzzy authentication at the block level. Due to the errors being localized at the column level, extra processing for iterative error correction is required. To improve the error localization, another algorithm called the image error correcting-noise tolerant message authentication code (IEC-NTMAC) is introduced. IEC-NTMAC is based on the EC-NTMAC algorithm [15] and marks the image blocks as suspicious, as opposed to the columns of image blocks, which is done by IEC-MAC.

Algorithm: IECC-MAC Tag Verification

Input:

- Received Compressed Image I′
- Received IECC-MAC′
- Image width W and height H
- Block length m
- Block LLRs: blockLLRs

Algorithm:

blocks′ = SplitInBlocks(I′, W, H, m)
unauthentic_columns = []
dcs_columns′$_i$_LLRs = BlockLLRsToDCColumnLLRs(blockLLRs)
authentic = true

for i = 1 to W/m
    dcs_column$_i$ = []
    for j = 1 to H/m
        dct = ComputeDCTOnBlock(blocks′$_{i,j}$)
        dcs_column′$_i$ = dcs_column′$_i$ || dct$_{1,1}$
    end
    C-MAC″$_i$ = ComputeMAC(k$_1$, dcs_column′$_i$)
        if(HD(C-MAC′$_i$, C-MAC″$_i$) ≤ d)
            status = PerformErrorCorrection(i, j, dcs_column′$_i$, dcs_columns′$_i$_LLRs)
            if(status == decoding_failure)
                authentic = false
                unauthentic_columns = unauthentic_columns || i
            end
        end
end

Output:

If (authentic == true)
        verification_status = true
else
        verification_status = false
        return unauthentic_columns
end

5 Fuzzy Image Authentication with Error Localization and Correction

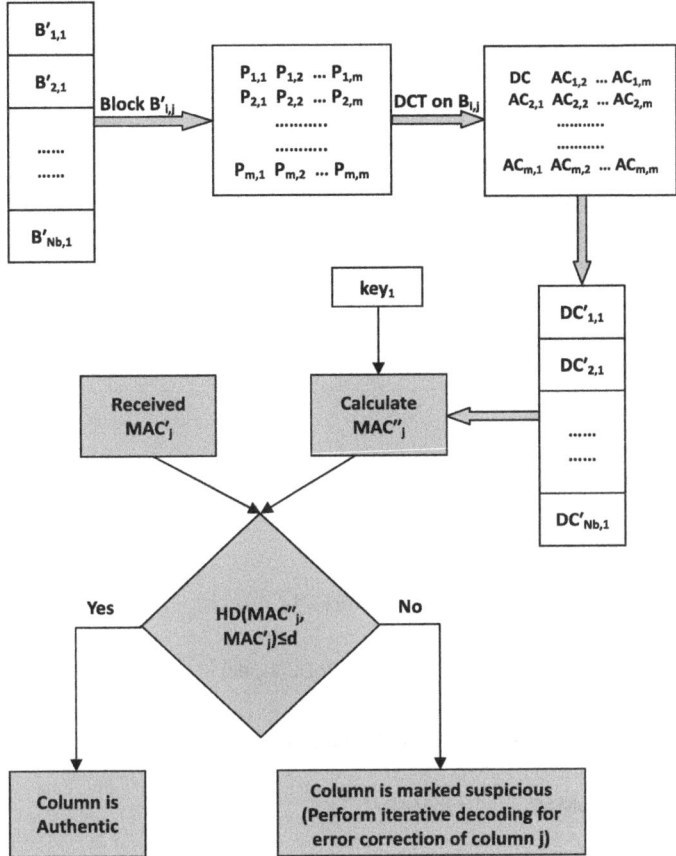

Fig. 5.3 Error localization and correction using IECC-MAC

5.4.3 Image Error Correction Noise Tolerant Message Authentication Code (IEC-NTMAC)

For error localization to a more fine grained level, an algorithm based on the EC-NTMAC [15] is proposed. It has an enhanced error localization and image recovery capability as compared to IEC-MAC.

The IEC-NTMAC algorithm works by calculating an NTMAC for each of the DC component in the DCT sub-matrix. Although a complete standard MAC is calculated for each DC component, only a small portion (called sub-MAC) is retained, e.g., the last s-bits of the MAC (see NTMAC and W-NTMAC [15]). All of these sub-MACs are appended together to constitute a complete NTMAC for each column of the DC matrix. Now the same step is repeated for all the columns, giving N/m NTMACs. Using an NTMAC not only improves the error localization to block level but also improves the error correction. The reason is that smaller blocks can be corrected more efficiently (in terms of processing power and memory consumption) as compared to the whole column of blocks.

IEC-NTMAC generation for an image I is shown in Fig. 5.4.

Algorithm: IEC-NTMAC Tag Generation

Input:

- Image I
- Image width W
- Image height H
- Block length m

Algorithm:

blocks = SplitInBlocks(I, W, H, m)

for i = 1 to W/m
    for j = 1 to H/m
        dct = ComputeDCTOnBlock(blocks$_{i,j}$)
        dc = dct$_{1,1}$
        subMAC$_{DCj}$ = ComputeSubMACOnDC(k_2, dc)
    end
    NTMAC$_{i,DC}$ = subMAC$_{DC1}$ || subMAC$_{DC2}$ || .... || subMAC$_{DCH/m}$
end
NTMAC$_{DC}$ = subMAC$_{1,DC}$ || subMAC$_{2,DC}$ || .... || subMAC$_{W/m,DC}$

Output:

NTMAC$_{DC}$

The receiver receives a compressed image (denoted as I′) as well as its IEC-NTMAC (denoted as IEC-NTMAC′). The receiver recalculates the IEC-NTMAC on I′ by using the same algorithm as explained above (let the recalculated IEC-NTMAC be denoted as IEC-NTMAC″). IEC-NTMAC′ is compared with IEC-NTMAC″ through a comparison of the corresponding sub-MACs. If the sub-MACs are d-fuzzy-authenticated according to definition 1, then the DC component is accepted as authentic and the image block corresponding to the DC component is declared authentic. Otherwise, if the sub-MACs are not d-fuzzy-authenticated, the block is marked as suspicious. All the blocks marked as suspicious identify the potential modifications in the image and are tried to be corrected using the iterative error correction algorithm explained in Sect. 2.2 on IEC-MAC. The major difference is that in Sect. 2.2, the whole suspicious column of image blocks is considered. Thus, it is good to identify the error locations to a column level but depending on the image resolution, it might be computationally very expensive to do the error recovery. In IEC-NTMAC, only the suspicious

5 Fuzzy Image Authentication with Error Localization and Correction

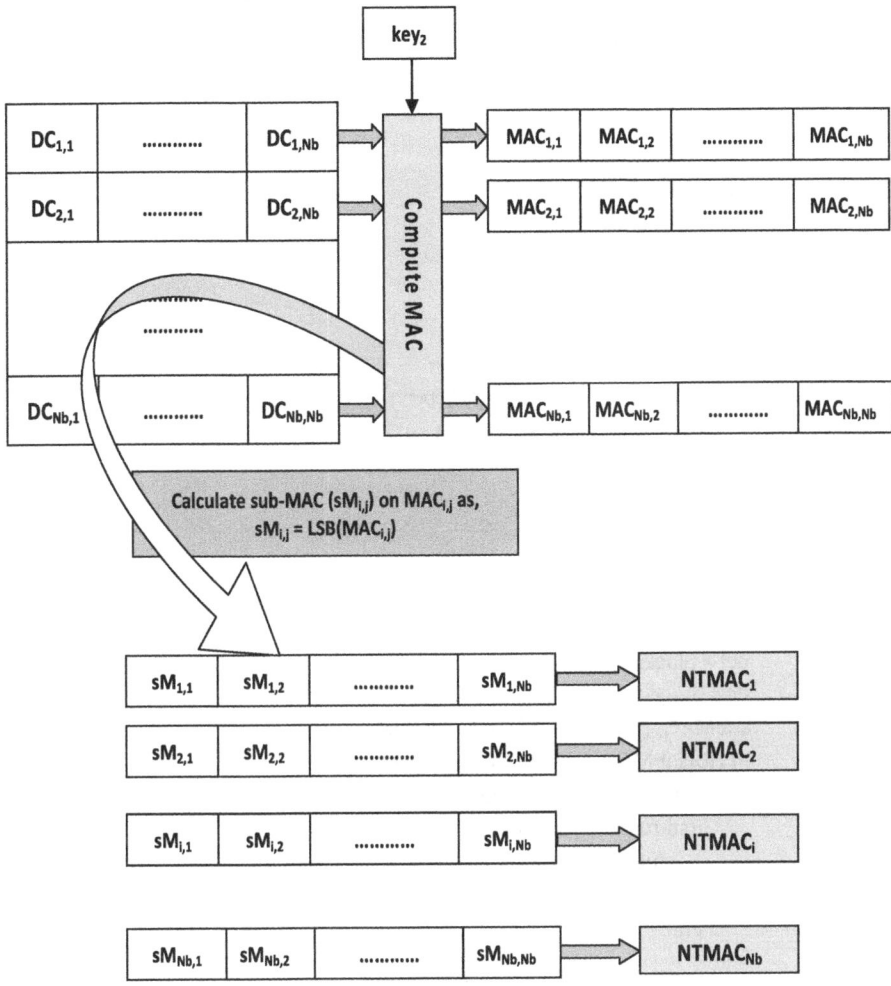

Fig. 5.4 Calculation of IEC-NTMAC at the sender side.

blocks are tried for error recovery, therefore the error recovery is computationally less expensive as compared to IECC-MAC. The pseudo-code of the IEC-NTMAC verification and error localization with error recovery at the receiver is given below and also depicted in Fig. 5.5.

Algorithm: IEC-NTMAC Tag Verification

Input:

- Received Compressed Image I′
- Received IEC-NTMAC′
- Image width W and height H
- Block length m
- Block LLRs: blockLLRs

Algorithm:

I″ = DecompressImage(I′)
subMAC′$_{DC}$ = MakeSubMACsForDCs(NTMAC′)
dc_LLRs = BlockLLRsToDCLLRs(blockLLRs)
blocks = SplitInBlocks(I″, W, H, m)
authentic = true
unauthentic_blocks = []

for i=1 to W/m
    for j=1 to H/m
        dct = blocks$_{i,j}$
        dc = dct$_{1,1}$
        subMAC$_{DCi,j}$ = ComputeSubMAC(k$_2$, dc)
        if(HD(subMAC′$_{DCi,j}$, subMAC$_{DCi,j}$) ≤ d)
            status = PerformErrorCorrection(i, j, dc, dc_LLRs$_{i,j}$)
            if(status == decoding_failure)
                authentic = false
                unauthentic_blocks = unauthentic_blocks || block$_{i,j}$
            end
        end
    end
end

Output:

If (authentic == true)
    verification_status = true
else
    verification_status = false
    return unauthentic_blocks
end

5 Fuzzy Image Authentication with Error Localization and Correction

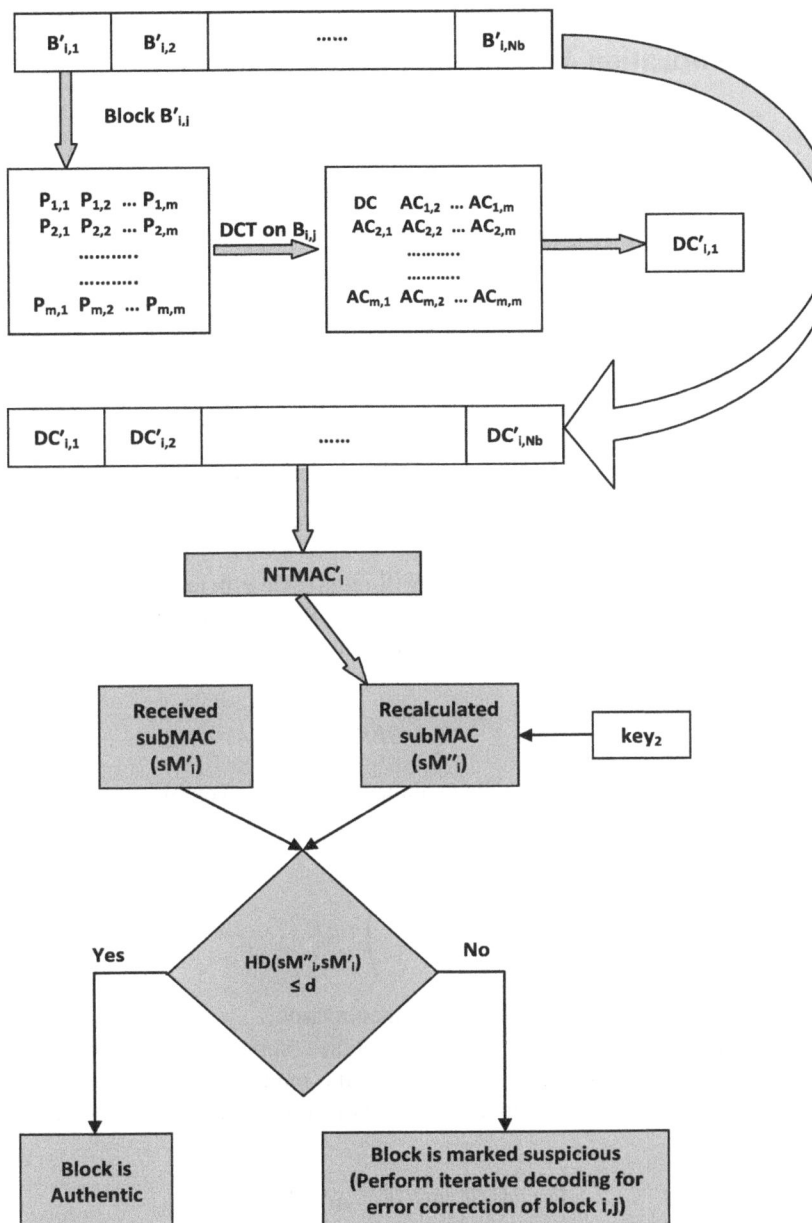

Fig. 5.5 Error localization and correction using IEC-NTMAC at the receiver side

5.5 Performance and Security Analysis of the Proposed Fuzzy Authentication Algorithms

5.5.1 Performance Study

To analyze the algorithms, it is assumed that an ideal n-bit MAC (n $\geq$ 256) is used for each column of the DC coefficients. The image is divided into $m \times m$ nonoverlapping blocks to get a total of m^2, k-bit DC coefficients.

At the receiver, the fuzzy authentication is considered to be successful if the Hamming distance between the received and the recalculated MACs does not exceed a threshold value. For such a fuzzy authentication algorithm, there is a possibility of the following two types of errors,

False Rejection A correct image (or a complete column of blocks or individual blocks) is discarded, though it should have been declared authentic.

False Acceptance An incorrect image (or a complete column of blocks or individual blocks) is accepted, though it should have been declared unauthentic.

If the BER is too high, false rejection will reduce the efficiency of the proposed schemes. When channel is in good state, i.e., the BER is low, false rejection will happen rarely. The probability of a false rejection for hard verification, i.e., with d = 0, is given by,

$$1 - (1 - BER)^n \tag{5.7}$$

The false rejection probability for fuzzy authentication is given by,

$$\Pr(FALSE\ REJECTION)$$
$$= (1 - BER)^{mk} \sum_{i=d+1}^{mn} \binom{mn}{i} BER^i (1 - BER)^{mn-i} \tag{5.8}$$

which is much smaller than that for hard authentication.

To calculate the probability of a false acceptance, suppose that R is the number of non-authentic column-wise MACs and E is the number of erroneous bits. The probability of a false acceptance in the presence of e erroneous bits in DC coefficient is given by,

$$\Pr(FALSE\ ACCEPTANCE)$$
$$= \Pr(R = 0 \mid E = e)$$
$$= \sum_{i=1}^{e} p_i q_i \tag{5.9}$$

where p_i is the probability of a false acceptance in each column when i DC coefficients are erroneous and q_i is the conditional probability of i erroneous DC coefficients when e bits are in error. p_i is approximated by binomial distribution while q_i can be estimated using classical occupancy problem [12–15]:

Pr (*FALSE ACCEPTANCE*)

$$= \sum_{i=0}^{e} \left(\frac{\sum_{k=0}^{d} \binom{m}{k}}{2^n} \right)^i \binom{m}{i} \sum_{j=0}^{i} (-1)^j \frac{\binom{i}{j}(i-j)^e}{m^e} \qquad (5.10)$$

5.5.2 Security Analysis

Algorithms introduced in this chapter are based on the ideal standard MAC, so the generic attacks on the standard MACs are considered as potential threats. In the given approaches, MACs may tolerate a modest number of errors; therefore, the security strength is reduced in general as compared to hard authentication MACs. In IECC-MAC, each DC element is protected by one MAC and the attacker has to forge the DC coefficients in such a way that column MACs can be d-fuzzy-authenticated. In IEC-NTMAC, the attacker even has more difficult task due to random partitioning [15].

A common approach to approximate the required complexity (data/time) for forgery attack on the MACs is given by "birthday paradox" which is based on finding collisions. For the fuzzy authentication, an attacker has to perform a near collision attack [37]. Near collision refers to a message pair, such that their MACs differ a little from each other. By extending the ordinary birthday paradox to the introduced fuzzy authentication scheme with a threshold d, it is expected to have a near collision with at most d-bit differences with the data complexity of,

$$\sqrt{\frac{2^n}{\sum_{d=0}^{d} \binom{n}{d}}} \qquad (5.11)$$

The value of n is usually chosen to be 256 (bits). The threshold value is set in such a way that the false acceptance and false rejection rates are minimized. There are other experimental methods to find a conveniently safe threshold zone by image processing techniques [10]. It can be observed that with the smaller threshold values, the security strength can be compensated by selecting longer MAC lengths [38]. The security of IEC-NTMAC is even higher due to secret partitioning. The attacker requires knowledge of partitioning methods before launching any attack.

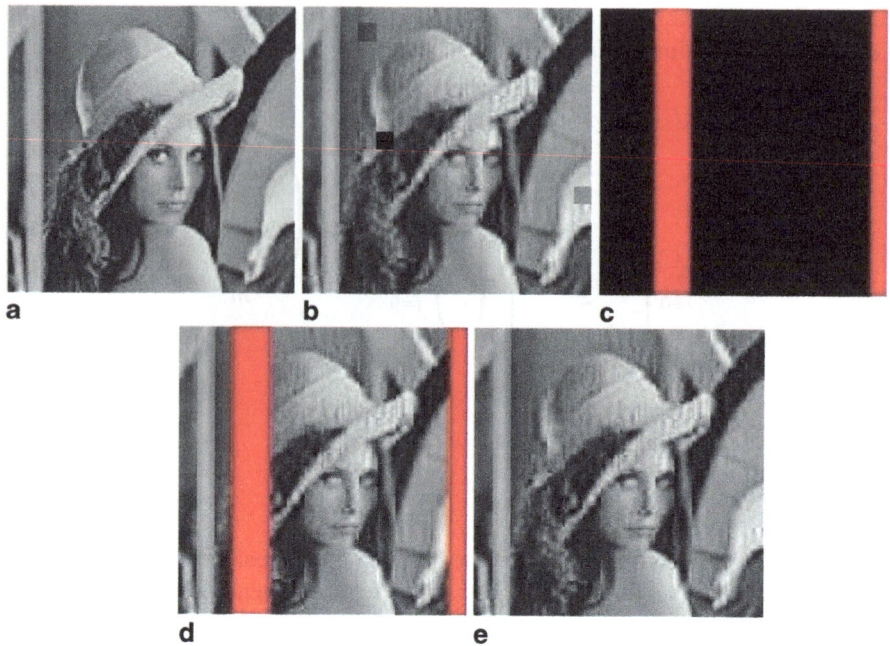

Fig. 5.6 IECC-MAC at 12.5 dB without channel coding.

5.6 Simulation Results

5.6.1 Simulation Parameters

Simulation results, for both of the proposed algorithms, are based on image transmission over additive white Gaussian noise (AWGN) channel with binary phase shift keying (BPSK) modulation. Results are given in the presence of as well as in the absence of channel coding (Turbo codes of rate-1/3). The LLRs produced by the decoder for convolutional turbo codes (CTC) are used for bit reliabilities, whereas the channel measurements are used as bit reliabilities in the absence of channel coding.

The source image is chosen as a grayscale image of 128×128 pixels, where each pixel requires 1 byte. The image is split into 8×8 pixel nonoverlapping blocks resulting in a total of 16×16 blocks for the chosen image resolution. DCT for each of these blocks is calculated and the DC components of the DCT matrices are protected using the algorithms explained earlier in Sect. 5.2. The DC and AC elements in the DCT matrix are of 16 bits each (and therefore need 2 bytes).

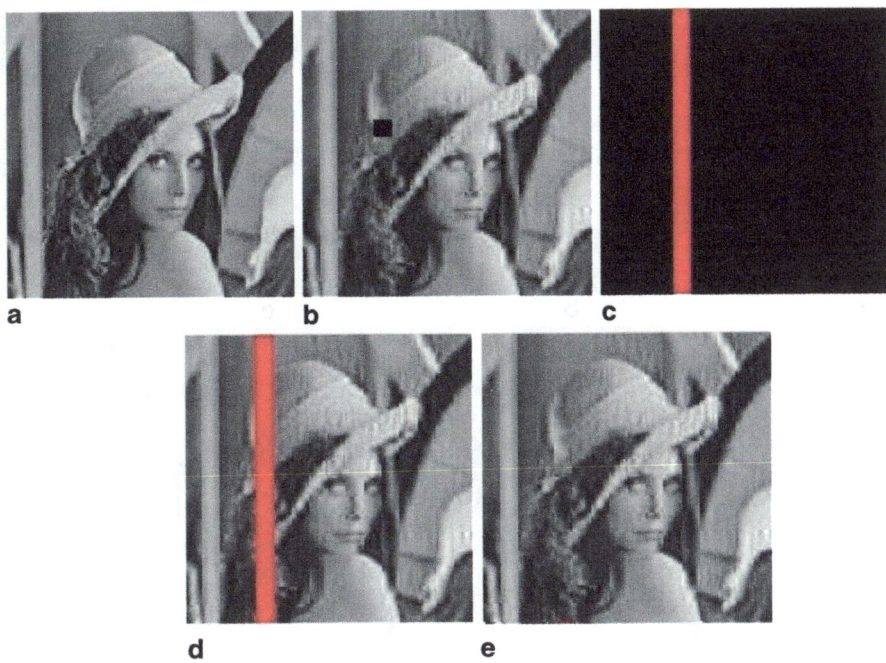

Fig. 5.7 IECC-MAC at 2.0 dB with CTC of rate-1/3

Fig. 5.8 a An airplane. **b** The number removed. **c** Modifications identified using IECC-MAC

5.6.2 Data Rate Analysis

SHA-256 is used as the standard MAC in the simulations. For a grayscale 128 × 128 pixels image protected using SHA-256, in total 128 × 128 × 8 bits of image data plus 256 bits of MAC (SHA-256) need to be transmitted. This is equal to 131,328 bits.

By protecting the DC elements using an IECC-MAC, 16 × 256 bits (= 4096 bits) of MACs are computed column-wise. The first 5 minor diagonals of the

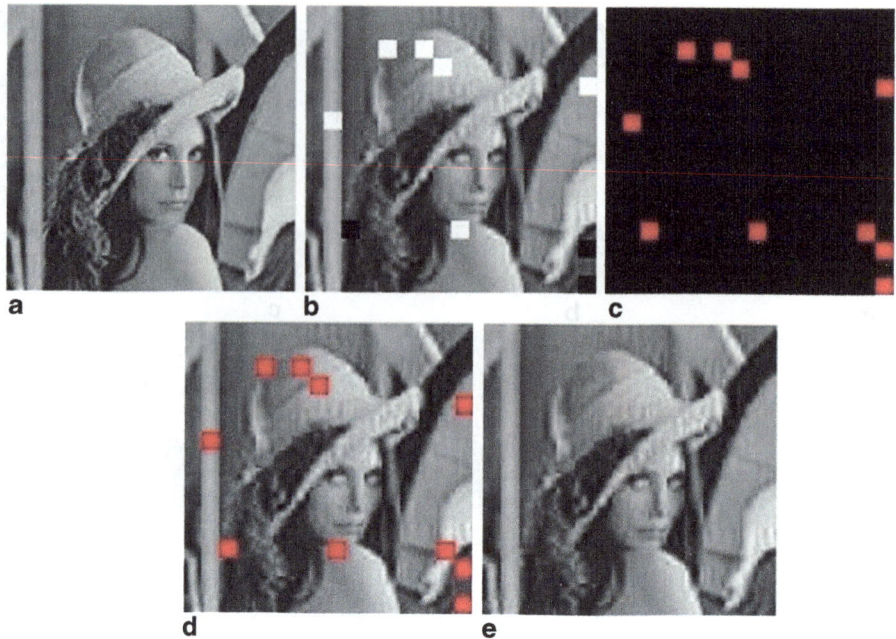

Fig. 5.9 IEC-NTMAC at 12 dB without channel coding

DCT matrices are transmitted (as compressed image in the simulations), which is $15 \times 16 \times 16 \times 16 = 61{,}440$ bits of data, where 15 is the number of elements from the 5 diagonals of each DCT matrix, 16×16 is the number of DCT matrices and the last 16 is the size (in bits) of each element in the DCT matrix. Thus, $61440 + 4096 = 65536$ bits are transmitted in total, which is approximately 50 % of the data that was transmitted by protecting the image using a standard MAC. Using IEC-NTMAC based error protection, the number of data bits transmitted is the same as in IECC-MAC. Each EC-NTMAC is 256 bits long. Thus, again approximately 50 % of the whole data is transmitted as compared to the conventional data transmission.

5.6.3 Simulation Results for IECC-MAC

Figure 5.6 shows the simulations results for IECC-MAC in the absence of channel coding at E_b/N_0 of 12.5 dB. Figure 5.6a shows the source image that is to be transmitted after protection with IECC-MAC. Figure 5.6b shows the image received at the receiver side. Figure 5.6c shows how the suspicious block positions are highlighted in red (columns of suspicious blocks) followed by the suspicious blocks mapped into the received image. Finally, the resultant image is shown, which is obtained by

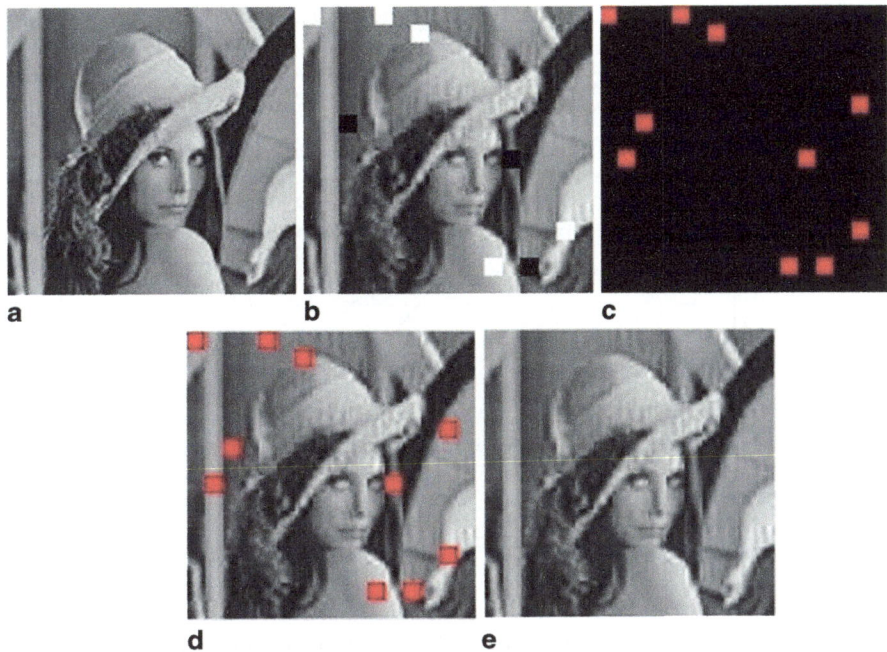

Fig. 5.10 a IEC-NTMAC at 2.75 dB with CTC of rate-1/3

Fig. 5.11 a An airplane. b The number removed (forgery). c Modifications identified using IEC-NTMAC

performing the error correction of IECC-MAC algorithms over the suspicious image based on the localized errors.

Figure 5.7 shows the simulation results for IECC-MAC in the presence of channel coding (Turbo codes of rate-1/3) at E_b/N_0 of 2.0 dB. It can be observed that due to the presence of turbo codes, lesser number of erroneous blocks is obtained at a much lower E_b/N_0.

Figure 5.8 shows the results of IECC-MAC after forgery attack. Figure 5.8a shows an airplane image. Figure 5.8b shows a forged image, where the serial number

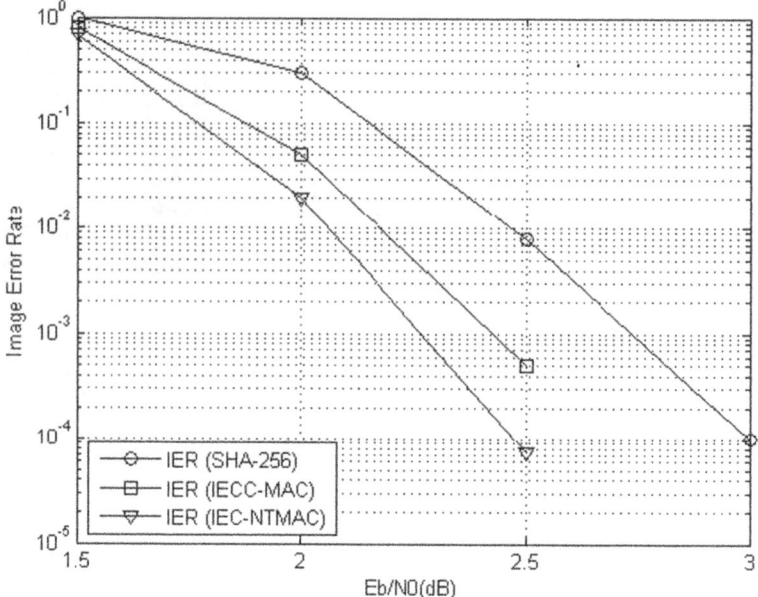

Fig. 5.12 IER over the AWGN channel with BPSK modulation and turbo codes of rate-1/3.

"01568" of the airplane has been removed. IECC-NTMAC is able to localize these modifications, as shown in Fig. 5.8c but is not able to correct them.

5.6.4 Simulation Results Using IEC-NTMAC

Images protected using IEC-NTMACs have better error localization capabilities and so they can be reconstructed more efficiently as compared to IECC-MAC. Simulation results in the presence and absence of channel coding, similar to the previous subsection, are presented in Fig. 5.9 and 5.10. Figure 5.11 shows the forgery attack on the airplane image. The error localization of IEC-NTMAC is refined to the block level, so it can be seen that the exact area of forgery has been marked by the IEC-NTMAC.

5.6.5 Image Error Rate (IER)

IER is defined as the number of times the whole image is declared as unauthentic divided by the number of times the image was received at the receiver, i.e.,

$$IER = \frac{\text{Nunmer of times the image is declared unauthentic}}{\text{Number of times the image is received}} \quad (5.12)$$

IER for both the algorithms is shown in Fig. 5.5 at different values of E_b/N_0. Three different curves are plotted for comparison. These curves show the IER in the presence of a standard MAC based protection, then in the presence of IECC-MAC and finally in the presence of IEC-NTMAC. Figure 5.5 shows that IEC-NTMAC provides the lowest IER and performs the best amongst the proposed algorithms (Fig. 5.12).

References

1. Reed IS, Solomon G. Polynomial codes over certain finite fields. SIAM J Appl Math. 1960;8:300–4.
2. Berrou C, Glavieux A, Thitimajshima P. Near shannon limit error-correcting coding and decoding: turbo-codes. Proceedings of the IEEE International Conference on Communications (ICC'93); Geneva, Switzerland; May 1993. pp. 1064–70.
3. Gallager RG. Low density parity check codes. Monograph, Cambridge: M.I.T. Press; 1963.
4. Peterson LL, Davie BS. Computer networks: a systems approach. 5th edn. Burlington: Morgan Kaufman; 2011.
5. Peterson WW, Brown DT. Cyclic codes for error detection. Proc IRE. 1961;49(1):228–35; (January 1961). doi:10.1109/JRPROC.1961.287814.
6. ISO/IEC 9797-1. Information technology—security techniques—Message Authentication Codes (MACs)—Part 1: mechanisms using a block cipher; 2011.
7. ISO/IEC 9797-2. Information technology—security techniques—Message Authentication Codes (MACs)—Part 2: mechanisms using a dedicated hash-function; 2011.
8. Bellare M, Canetti R, Krawczyk H. Keying hash functions for message authentication. Advance in Cryptology-CRYPTO'96, Lecture Notes in Computer Science, vol. 1109, N. Koblitz ed., Springer-Verlag, 1996. pp. 1–15.
9. Ur-Rehman O. Applications of iterative soft decision decoding. Aachen: Shaker Verlag; 2013. ISBN:978-3-8440-1641-3.
10. Graveman R, Fu K. Approximate message authentication codes. Proceedings of 3rd Annual Fed lab Symposium on Advanced Telecommunications/Information Distribution, vol. 1, College Park, MD, Feb. 1999.
11. Gravemen R, Xie L, Arce GR. Approximate image message authentication codes. Proceedings of 4th Annual Symposium on Advanced Telecommunications and Information Distribution Research Program, College Park, MD; 2000.
12. Boncelet C. The NTMAC for authentication of noisy messages. IEEE Trans Inf Forensics Secur. 2006;1(1):35–42.
13. Onien D, Safavi-Naini R, Nickolas P, Desmedt Y. Unconditionally secure approximate message authentication. Proceedings of the Second International Workshop on Coding and Cryptology, Springer, 2009.
14. Ge R, Arce GR, Crescenzo GD. Approximate message authentication codes for N-ary alphabets. IEEE Trans Inf Forensics Secur. 2006;1(1):56–67.
15. Ur-Rehman O, Zivic N, Amir Hossein S, Tabatabaei AE, Ruland C. Error correcting and weighted noise tolerant message authentication codes. 5th International Conference on Signal Processing and Communication Systems (ICSPCS)/IEEE Conference, Hawaii, USA, December; 2011.
16. Zivic N, Flanagan M. On joint cryptographic verification and channel decoding via the maximum likelihood criterion. IEEE Commun Lett. 2012;6(5):717–9.
17. Zivic N. Joint channel coding and cryptography. Aachen: Shaker Verlag; 2008. ISBN:978-3-8322-7180-0.
18. Ur-Rehman O, Tabatabaei AE, Amir Hossein S, Zivic N, Ruland C. Soft authentication and correction of images. Systems, Communication and Coding (SCC), Proceedings of 2013 9th International ITG Conference on, 2013; pp. 1–6.

19. Ur-Rehman O, Zivic N. Noise tolerant image authentication with error localization and correction. Communication, Control, and Computing (Allerton), 2012 50th Annual Allerton Conference on, 2012; pp. 2081–7.
20. Chen WH, Pratt WK. Scene adaptive coder. IEEE Trans Commun. 1984;COM-32:225–32.
21. Watson A. Image compression using discrete cosine transform. Math J. 1994;1(4):81–8.
22. Ahmed N, Natarajan T, Rao KR. Discrete cosine transform. IEEE Trans Comput. 1974;C-23:90–3.
23. Yip P, Rao KR. Fast decimation-in-time algorithms for a family of discrete sine and cosine transforms. Circuits Syst Signal Process. 1984;3:387–408.
24. Chung SY, Forney GD Jr, Richardson TJ, Urbanke R. On the design of low-density parity-check codes within 0.0045 dB of the shannon limit. IEEE Commun Lett. 2001;5(2):58–60.
25. Berlekamp ER. Algebraic coding theory. New York: McGraw-Hill; 1968. (Revised edition, Laguna Hills: Aegean Park Press, 1984).
26. Sugiyama Y, Kasahara Y, Hirasawa S, Namekawa T. A method for solving key equation for goppa codes. Inf Control. 1975;27:87–99.
27. Guruswami V, Sudan M. Improved decoding of reed-solomon and algebraic-geometry codes. IEEE Trans Inf Theory. 1999;45(6):1757–67.
28. Koetter R, Vardy A. Algebraic soft-decision decoding of reed-solomon codes. Proceedings of the 2000 IEEE International Symposium on Information Theory, p. 61, Sorrento, Italy, Jun. 25–30, 2000.
29. Elias P. List decoding for noisy channels. Technical Report 335, Research Laboratory of Electronics, MIT; 1957.
30. Cox I, Miller M, Bloom J, Fridrich J, Kalker T. Digital watermarking and steganography. Berlington: Morgan Kaufmann; 2007.
31. Lee J, Won CS. A watermarking sequence using parities of error control coding for image authentication and correction. IEEE Trans Consumer Electron. 2000;46(2):313–7.
32. Wu Y. Tamper-localization watermarking with systematic error correcting code. Proceedings of IEEE International Conference on Image Processing (ICIP). 2006; pp. 1965–8, Atlanta, GA.
33. Fridrich J, Goljan M. Images with self-correcting capabilities. Proceedings of the IEEE International Conference on Image Processing (ICIP). 1999; pp. 792–6, Kobe.
34. Umbaugh SE. Digital image processing and analysis: human and computer vision applications with CVIPtools. 2nd edn, CRC Press, Nov. 2010. ISBN:978-1-4398-0205-2.
35. International Organization for Standardization. ISO/IEC JTC 1/SC 29 (2009-05-07). ISO/IEC JTC 1/SC 29/WG 1—coding of still pictures (SC 29/WG 1 Structure); 2009.
36. International Organization for Standardization. ISO/IEC 15444-1:2004—information technology—JPEG 2000 image coding system: core coding system; 2004.
37. Preneel B, van Oorschot PC. MDx-MAC and building fast MACs, from hash functions. Proceeding of CRYPTO 1995, LNCS 963, Springer Verlag. 1995; pp. 1–14.
38. Zivic N. Coding and cryptography: synergy for robust communication. Munich: Oldenbourg Verlag; 2013. ISBN:978-3-486-75212-0.

Chapter 6
Robustness of Biometrics by Image Processing Technology

Robin Fay and Christoph Ruland

6.1 Introduction

Authentication is a crucial process to authorize users, computers, or other entities in an interconnected society. Passwords or tokens are often used to authenticate users. However, passwords can be forgotten and tokens get lost. Biometric traits like fingerprints or vein patterns are always present, differ from one person to another, and can also be used for user authentication. Moreover, they cannot be forgotten or lost. Biometric authentication has found its way into everyday life through smartphones and notebooks. Especially due to the wide use of biometric authentication the protection of biometric user information has become inevitable because, unlike passwords, this information cannot be changed in case of compromising.

One major subject of a computer aided biometric system is the extraction of meaningful features from captured images. These features should represent the uniqueness of the object to ensure a correct classification. Differences between the images, for instance caused by rotation or translation of the finger during fingerprint acquisition or noisy image data, are leading to classification errors. Therefore, suitable preprocessing techniques are needed. In general, noise will be reduced using appropriate image enhancement, and image transformations can be corrected through image alignment. However, alignment is a computational expensive operation, and its robustness is crucial in regards to classification performance. Furthermore, the private biometric user information could be attacked by exploiting the additional alignment information. Therefore, this chapter focuses on alignment-free feature extraction methods for the usage in biometric systems with template protection.

R. Fay (✉) · C. Ruland
Chair for Data Communications Systems, University of Siegen, 57076, Siegen, Germany
e-mail: robin.fay@uni-siegen.de

C. Ruland
e-mail: christoph.ruland@uni-siegen.de

This chapter is organized as follows: Sect. 6.2.1 gives a brief introduction to the general pattern recognition process. Here, a special focus lies on affine invariant pattern recognition and alignment-free features. In Sect. 6.2.3 biometric authentication systems and template protection is described followed by fingerprint and vein pattern recognition. The basic ideas of different feature extraction approaches are presented in Sect. 6.3, before some proposed systems which are described in Sect. 6.4. An evaluation strategy for local feature extraction methods based on criteria from the field of digital image processing is shown in Sect. 6.5. Finally, a conclusion and some ideas for further research are depicted in Sect. 6.6.

6.2 Pattern Recognition and Biometric Authentication

6.2.1 Pattern Recognition Systems

Assume that V is the set of all possible patterns f, which are commonly images. Each pattern f should have a desired classification $c \in \Omega$ where Ω is the set of all classes. The goal of a *pattern recognition system* is to determine a function $h : V \to \Omega$ that finds the desired classification c for all $f \in V$ [1]. A pattern recognition system consists of five stages as shown in Fig. 6.1.

Acquisition Patterns are captured by a sensor, for example, a camera taking an image. The main properties of the captured pattern arise from technical characteristics of the sensor, like the type (e.g., optical or capacitive), resolution, or color scheme.

Preprocessing After the acquisition, the pattern needs to be preprocessed. In case of an image, for example, appropriate preprocessing techniques can be image enhancement for noise reduction or alignment to deal with perspective transformations. Proper preprocessing depends strongly on the pattern itself and the chosen feature extraction technique. Different kinds of preprocessing methods from the domain of computer vision are presented in [2].

Feature Extraction During feature extraction, the objective is to reduce the pattern to meaningful feature x for classification instead of processing the whole pattern. In machine-based pattern recognition, a feature $x \in \mathbb{R}$ is a measurement constructed

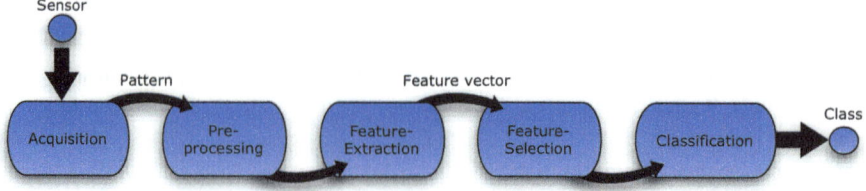

Fig. 6.1 The general pattern recognition pipeline consisting of five stages

from a particular pattern that makes this pattern comparable or distinguishable. In other words, the features should represent the significant information of the pattern with the aim to characterize it uniquely. This approach is similar to the human visual perception, since perception depends only on a small amount of relevant characteristics [3].

The result of feature extraction is a n-dimensional feature vector x = $[x_1, x_2, \ldots, x_n]^t$ that contains all features from the pattern and therefore clearly describes it. Automatic feature extraction is the most important part in a pattern recognition system, since the classification depends only on the extracted features. Features can also be extracted manually, but this approach is not suitable for huge pattern databases. Feature extraction methods for fingerprint and vein biometrics are considered in Sects. 6.3 and 6.4.

Feature Selection Assuming that the extracted feature vector contains spurious features, feature selection can be used to reduce the amount of these unintentionally extracted features, which also reduces the dimension of the feature vector. Spurious features can be redundant features, which carry no additional information with respect to particular other features, and insignificant features carrying little or no useful information at all. Both should be avoided, because these features are not suitable to distinguish between the patterns. See a detailed description of feature selection techniques in [4] and [5].

Classification Classification is the last stage of pattern recognition. The classification process is a kind of supervised learning, where the classifier is trained according to a labeled training set to determine the desired classification. A lot of classification techniques were proposed, for example, artificial neural network [6], support vector machines [7], or others [8].

6.2.2 Affine Invariant Pattern Recognition

6.2.2.1 The Problem of Affine Invariant Pattern Recognition

As already illustrated, a serious problem of computer vision is the recognition of shifted or rotated images. If a pattern inside the training set belongs to class c, the same pattern, with a change in orientation, should also be assigned to class c. This leads to the problem of affine invariant pattern recognition. First consider that a function is called invariant under a group of transformations, if a transformation of this group can be applied to the argument of the function without changing the functions value. Now let G be the group of all affine transformations, so that $g \in G$ is an affine transformation. A pattern recognition system is *invariant* under g if

$$h(gf) = h(f) \Leftrightarrow h(g\mathbf{x}) = h(\mathbf{x}) \qquad (6.1)$$

applies to all patterns $f \in V$ with the corresponding feature vector x [1].

Affine invariance can be achieved by finding the affine transformation g that aligns the patterns. In order to find the affine transformation g, the affine transformation matrix must be identified by either manual selection or automatic detection of corresponding points. Another approach is called alignment-free feature extraction, which is the extraction of invariant features, that directly leads to invariant classification. If **x** is invariant under g this also implies that $h(g\mathbf{x}) = h(\mathbf{x})$. Therefore, the characteristics and types of the features must be considered in detail.

6.2.2.2 Alignment-Free Features

Features can be separated into two types, namely *global* and *local* features. Global features describe the whole pattern like, for example, *image moments* which describe the whole image including background. Image moments are features that represent global image properties such as the centroid or the sum of all color values of each pixel.

For a better understanding of global features and how they can be constructed to become invariant against particular affine transformations, the theory of image moments will be considered briefly.

Image Moments

Let $f(x, y)$ be an image function of a discrete ($N \times M$) gray value image with $x = 0, 1, 2, \ldots (N - 1)$ and $y = 0, 1, 2, \ldots (M - 1)$ limiting the pixel position. *Discrete geometric moments* $\{m_{pq}\}$ of order $(p + q)$ are defined by [9]:

$$m_{pq} = \sum_{y=0}^{M-1}\sum_{x=0}^{N-1} x^p y^q f(x, y) \text{ with } p, q = 0, 1, 2, \ldots n \quad (6.2)$$

Take, for example, the discrete geometric moment of order zero to illustrate, what it describes in the image.

$$m_{00} = \sum_{y=0}^{M-1}\sum_{x=0}^{N-1} x^0 y^0 f(x, y) = \sum_{y=0}^{M-1}\sum_{x=0}^{N-1} f(x, y) \quad (6.3)$$

Considering the assumption that $f(x, y)$ is a discrete image function of a gray value image, it is easy to see that $\{m_{00}\}$ is the sum of all gray values of this image. With the help of zero and first order moments, the image centroid $(\bar{x}, \bar{y})$ can be determined by [10]:

$$\bar{x} = \frac{m_{10}}{m_{00}}, \bar{y} = \frac{m_{01}}{m_{00}} \quad (6.4)$$

Translation invariant moments can be defined with respect to the image centroid, since the image centroid is not changed by application of translations. This leads to the definition of translation invariant *central moments* $\{\mu_{pq}\}$ [11]:

$$\mu_{pq} = \sum_{y=0}^{M-1}\sum_{x=0}^{N-1} (x - \bar{x})^p (y - \bar{y})^q f(x, y) \quad (6.5)$$

To obtain the invariance against scale changes, one could normalize the images to the same size, because global features are constructed to describe the whole image. This idea can be adapted to get *normalized central moments* $\{\eta_{pq}\}$, which are defined by [11]:

$$\eta_{pq} = \frac{\mu_{pq}}{\mu_{00}^{\gamma}}, \text{ with } \gamma = \frac{(p+q)}{2} + 1 \qquad (6.6)$$

Hu [12] constructed a set of seven invariant moments $\{h_i\}$ from the normalized central moments, that are not affected by changes in translation, scale, and rotation. These seven invariant moments are defined by [11]:

$$h_1 = (\eta_{20} + \eta_{02}), \qquad (6.7)$$

$$h_2 = (\eta_{20} - \eta_{02})^2 + 4\eta_{11}^2,$$

$$h_3 = (\eta_{30} - 3\eta_{12})^2 + (3\eta_{21} - \eta_{03})^2,$$

$$h_4 = (\eta_{30} + \eta_{12})^2 + (\eta_{21} + \eta_{03})^2,$$

$$h_5 = (\eta_{30} - 3\eta_{12})(\eta_{30} + \eta_{12})[(\eta_{30} + \eta_{12})^2 - 3(\eta_{21} + \eta_{03})^2]$$
$$+ (3\eta_{21} - \eta_{03})(\eta_{21} + \eta_{03})[3(\eta_{30} + \eta_{12})^2 - (\eta_{21} + \eta_{03})^2],$$

$$h_6 = (\eta_{20} - \eta_{02})[(\eta_{30} + \eta_{12})^2 - (\eta_{21} + \eta_{03})^2]$$
$$+ 4\eta_{11}(\eta_{30} + \eta_{12})(\eta_{21} + \eta_{03}),$$

$$h_7 = (3\eta_{21} - \eta_{03})(\eta_{30} + \eta_{12})[(\eta_{30} + \eta_{12})^2 - 3(\eta_{21} + \eta_{03})^2]$$
$$- (\eta_{30} + 3\eta_{12})(\eta_{21} + \eta_{03})[3(\eta_{30} + \eta_{12})^2 - (\eta_{21} + \eta_{03})^2].$$

An example of these moments is illustrated in Fig. 6.2 which also shows the invariance. A logarithmic representation is often suggested due to the low numerical value of the invariant moments [13]:

$$h_{log,i} = |log(|h_i|)| \qquad (6.8)$$

This is only one example of global features. Other examples are, for instance, *Legendre moments* [14], *Zernike moments* [15], and *pseudo-Zernike moments* [9].

One disadvantage of global features is that local image changes affect all features and therefore the whole feature vector, which leads to misclassification. To counteract this problem, *subdividing* (also called *segmentation*) methods can be used, which divide the image into a certain number of sub-images. After that, global features can be generated for all particular images which increase the number of features, and reduce the influence of local image changes to a subset of the feature vector. The term subdividing should be used to describe the process of dividing images, since segmentation in the field of biometrics often means to separate the region of interest from the background, for example, the finger in a finger vein image.

An additional disadvantage of global features is that they cannot differentiate between the image background and other parts of the scene, so that the information is merged.

Fig. 6.2 (*left*) Original image. Invariant moments: $h_{log,1} = 0.5222$, $h_{log,2} = 2.1652$, $h_{log,3} = 4.5434$, $h_{log,4} = 3.5669$, $h_{log,5} = 7.6222$, $h_{log,6} = -4.8822$, $h_{log,7} = -9.1521$. (*right*) Image rotated by 45°. Invariant moments: $h_{log,1} = 0.5212$, $h_{log,2} = 2.1569$, $h_{log,3} = 4.5542$, $h_{log,4} = 3.5682$, $h_{log,5} = 7.6305$, $h_{log,6} = -4.8631$, $h_{log,7} = -8.7831$

Local Features

In comparison, local features describe the pattern locally through interesting keypoints or regions with respect to their direct neighborhood. Based on Schmid and Mohr [16] local feature extraction can be divided in two processes, namely *detection* and *description*, as shown in Fig. 6.3.

A feature detector identifies keypoints or regions at characteristic image positions. Typically, keypoints are points where the image signal changes intently, like corners or edges. The usage of interesting points should reduce the amount of data, that need to be examined, to the most significant information. For each interesting keypoint or region, a vector of local attributes—called descriptor—is computed to uniquely describe its local neighborhood. The feature vector consists of the descriptors of all keypoints. Consequently, a local feature is defined as a tuple of a point or region and a characteristic description of its surrounding, which can assume any desired shape (usually circles or ellipses are used).

Based on this definition, reliable local features should satisfy some additional properties. The first property arises from the advantage of local features against global features and is called *locality* [17]. Local image changes should affect only a part of the extracted features so that the resulting feature vector becomes more robust against local distortion. Local features should also represent only semantically significant regions with a high *information content* that clearly distinguish the represented pattern from any other.

In addition, features which are present in two images of the same scene or object should be detected in both images to obtain a similar feature vector. This property is called *repeatability* and is mainly influenced by the *invariance* of the feature detector against the occurring transformations [17]. Beyond this, repeatability is also influenced by *robustness*, for example, against noise, which can be either achieved by robust feature detection or by using appropriate preprocessing techniques.

Fig. 6.3 Local features—Detection and description. (Referring to [16]). First, interesting keypoints need to be detected before their neighborhood can be described

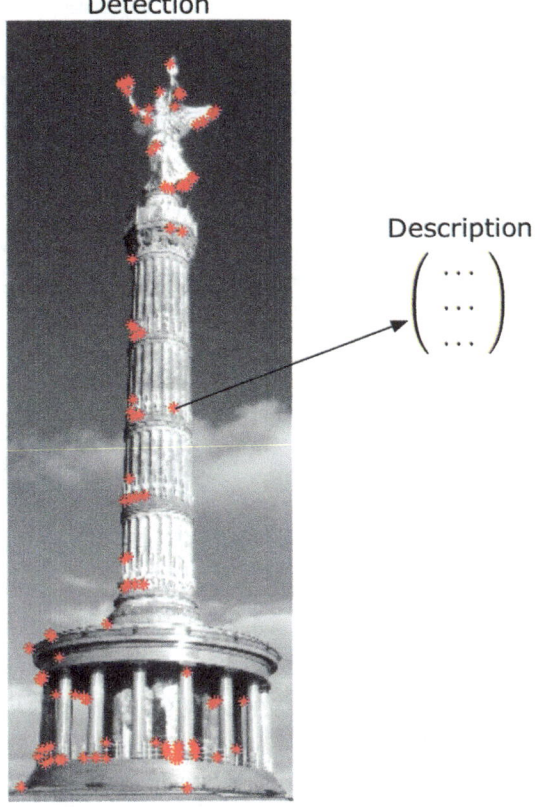

Summing up, it can be said that alignment-free local features can be achieved by invariant detection with high repeatability and also invariant description of interesting regions with a high information content. Examples of invariant feature detection can be found in [17] or [18], and different feature descriptors are described in [19]. Local alignment-free feature extraction methods proposed for fingerprint and vein authentication systems are considered in detail in Sect. 6.4.

6.2.3 Biometric Authentication

6.2.3.1 Biometric Systems

A biometric system is, in general, a pattern recognition system that operates on the basis of biometric data—precisely behavioral and physiological traits. A common process, similar to the pattern recognition pipeline from Fig. 6.1, can be used to describe biometric systems. The main components are sensor, preprocessing, feature extraction, feature selection, matching, and a system database (see Fig. 6.4).

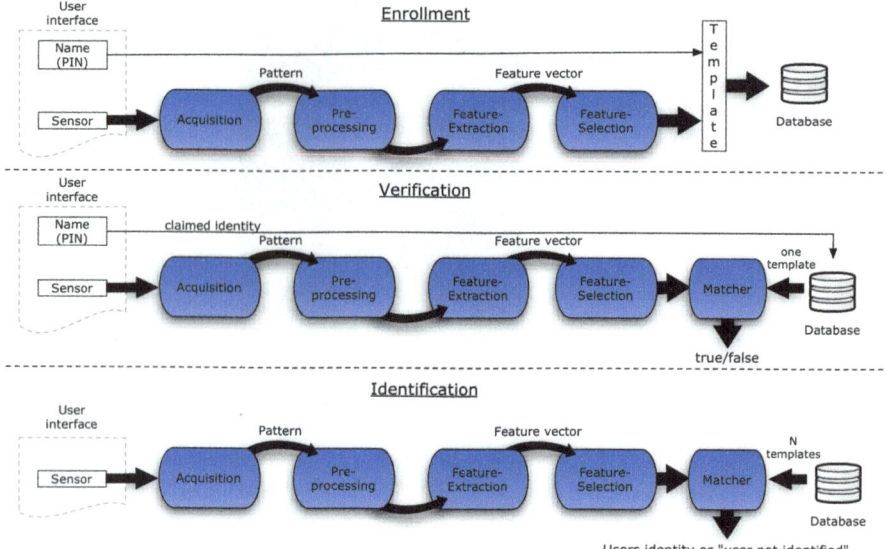

Fig. 6.4 Biometric systems as pattern recognition systems. (Referring to [20])

Furthermore, these systems can be divided into three modes, namely: *enrollment*, *verification*, and *identification*, which are displayed in Fig. 6.4 and explained hereinafter.

- **Enrollment:** The process, at which a user registers his traits in a biometric system, is called enrollment. First, a pattern is acquired using a biometric sensor. The quality of the pattern is conditioned and checked during the preprocessing step, so that recordings with poor quality can be repeated. A method for feature extraction is used to generate a feature vector, whose dimension is reduced by feature selection as mentioned in Sect. 6.2.1. The feature vector is stored in the system database together with personal information about the user and a personal identification number (PIN) as the user's *template* [21]. This template serves the biometric system as a basis for the verification and identification process. In addition, some systems offer to record different biometric traits for one user.
- **Verification:** A user demands a verification by entering his PIN or showing his identification card. In the verification mode, the system extracts the user's information first and compares it with the information stored in the biometric template, which belongs to the PIN or identification card. The system then takes the decision, if the user is the one he claims to be or not. The aim of verification is to prevent multiple users to share one identity [22].
- **Identification:** During identification, the acquired biometric features will be compared with all templates inside the database to find the current user under all known identities. Therefore, there is no need for the user to prove his/her identity by a PIN or identification card. One user with multiple identities or the authorization of a

6 Robustness of Biometrics by Image Processing Technology

non registered user will be prevented through the identification process. Biometric user identification is also used in forensic science.

Biometric systems are based on different biometric traits; an overview can be found in [22]. It is important to differentiate between the term feature as a significant measure, which distinguishes patterns, and the term biometric trait as a characteristic of an individual. This chapter focuses on fingerprint and vein pattern analysis, which will be considered in detail in the Sects. 6.2.3.3 and 6.2.3.4.

6.2.3.2 Template Protection

The usage of biometric authentication systems offers both advantages and further challenges. On the one hand, biometric traits are associated closely to the user and can not be forgotten or lost like, for instance, passwords. They are unique differentiators that always allow authentication. On the other hand, this close coupling is the reason for additional challenges. Beside dummy or violent attacks, there is another problem in the immutability of biometric traits. Once biometric traits are stored as a template, they are known to the system and cannot be changed. This raises the question what should happen if this data is compromised successfully. Passwords, for example, can be reset, but biometric traits are unchangeable.

If an unsecured channel is used to transfer the biometric data after acquisition, this offers a possibility for an eavesdropper to obtain the user's template. Since these data cannot be changed, an attacker has permanent access by injecting the monitored data, bypassing the sensor, as long as the affected user is not blocked by the system. After blocking, however, it is no longer possible for the user to authenticate himself. Therefore, biometric systems must be additionally protected against these attacks, and the biometric user templates must be protected particularly. It can be concluded that biometric templates may not be transmitted, stored, and processed without additional protection. To ensure the security of the biometric template, *protected templates* are used that should comply with the following properties [23]:

- **Irreversibility:** It should be impossible to obtain the originally recorded biometric data or any information derived from them with the help of the protected template.
- **Cancelable:** Protected templates should provide the opportunity to be revoked in case of a possible compromising.
- **Diversification:** It should be easily possible to produce different protected templates of the same or very similar biometric data.
- **Universal:** Protected templates are supposed to be generated for all kinds of biometric traits and also to be interoperable between different biometric systems. This needs uniform interchange standards for features, which describe biometric traits, like defined in ISO/IEC 19794 [24].

Biometric systems with template protection are often called *cancelable biometrics* [25]. Examples for cancelable biometrics can be found in [26, 27].

A very simple approach could be to apply a secret function on a feature vector to obtain a transformed feature vector, which does not contain the original information.

Fig. 6.5 A fingerprint image as a pattern of ridge lines (*dark*) and valleys (*lighter*)

In the case of a possible compromise of the data, one could easily change the function used for transformation. This basic idea shows a significant problem if image alignment is used. It is no longer possible to align the acquired image with respect to the protected template, because the only information of the original biometric traits is transformed features. To be able to use alignment anyway, additional alignment information is needed, and it is not clear, if this can be exploited to obtain the original template data. Hence, alignment-free features are required to get rid of the vulnerable alignment information.

6.2.3.3 Fingerprint Recognition

Fingerprints are probably the most commonly used biometric traits. A fingerprint is a pattern that is formed by the friction ridges of a fingertip skin [28]. This pattern consists of ridges and valleys, which may have various forms, for instance, arches, loops, or whorls. These characteristics, together with the interruptions of the friction ridges, called *minutia*e, make a fingerprint undoubtedly distinguishable [28]. At this point it is important to mention that the acquired image from the fingerprint is a gray value image as shown in Fig. 6.5. In order to see how features can be extracted from the fingerprint images, it should be considered which features are suitable for the usage in fingerprint recognition, and how they can be described.

Like in Sect. 6.2.2.2, features of fingerprints can be divided in global and local features.

Global Features Global fingerprint features are features that describe the whole pattern, like the five singularities: *left loop, right loop, whorl, arch*, and *tented arch* described by Henry [29]. These features are not applicable for computer aided fingerprint recognition because of their very limited information content. Global image features, such as moments, can also be considered as global features of fingerprints, since fingerprints are recorded as images.

Local Features On the local level, the most common features used in fingerprint recognition are minutiae points. Minutiae are small elements in the fingerprint, which are located at different types of interruptions or endings of ridge lines [30]. Examples of minutiae types are ridge endings, bifurcations, or intersections. Minutiae are extracted in the two steps (detection and description) as mentioned in Sect. 6.2.2.2, since minutiae are local features. Other than minutiae, fingerprints contain other significant attributes like ridge width and shape or pores, which can also be used to construct features [31]. Minutiae points and other significant fingerprint characteristics can be seen as a special case of keypoints like described in Sect. 6.2.2.2, as the image signal changes intently. Beside minutiae-based approaches, which specifically search for minutiae points and describe them, for instance, based on their position or type, image based methods are also used.

Image-based approaches are derived from the field of *digital image processing* and are originally used for object detection or general image classification. Interesting regions with a high information content are detected by finding keypoints and described using image based information. These techniques are also applied in biometric systems for fingerprint or vein recognition, see [32, 33] for examples. Image-based detectors also find minutiae points or pores to some extent, since minutiae are in general keypoints [34].

Furthermore, the number of general interesting points that arise from image-based detectors is between 500 and 3000, depending on image quality, preprocessing and other parameters [35]. However, the quantity of minutiae points is often smaller than 100 [32], and this number may decrease to less than 50 in the presence of noise or scars [35]. On the one hand, this means that in some cases it may not be possible to extract a sufficient number of features using minutiae points. On the other hand, if nearly the whole image is covered by keypoints the locality property and the advantages of local features are destroyed. Features from fingerprints should satisfy the properties of Sect. 6.2.2.2 such as locality, repeatability, and high information content. Therefore, invariance against translation and rotation of the finger during fingerprint acquisition is desired. Scale invariance may also be requested, if recordings from sensors with different resolutions are used.

6.2.3.4 Vein Pattern Recognition

Another biometric trait used for biometric authentication is vein patterns. These patterns are commonly acquired using near infrared light sources. This makes use of the fact that, near infrared radiation permeates tissue, but gets absorbed by the deoxygenated hemoglobin inside the blood [39]. The result is a gray value image of dark vein patterns as shown in Fig. 6.6.

Basically, all veins inside the human body can be used for authentication, but with respect to usability mostly hand, palm, or finger veins are utilized [40]. Vein patterns are suitable for biometric authentication, because they are universal, unique, and permanent. In addition, they cannot be revealed unintentionally, like for instance fingerprints left on a glass. Furthermore, vein patterns offer the possibility to detect,

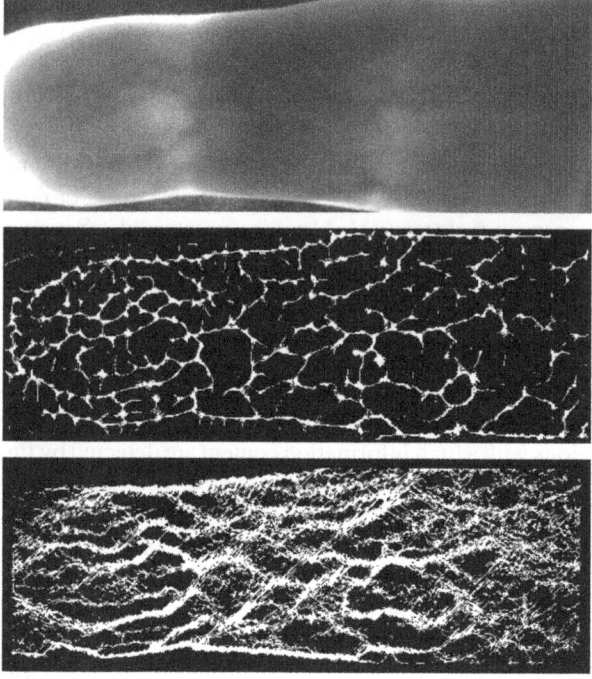

Fig. 6.6 An illustration of a finger vein image (*top*). The other images show examples for vein pattern extraction using the methods of [36] (*middle*) and [37] (*bottom*). (Referring to [38])

whether the user is alive during acquisition or not, which offers security against dummy attacks [39]. An interchange standard for vein images is defined in ISO 19794-9 [24].

Fingerprints and finger veins offer a possibility for multiple biometrics, which means the fusion of several biometric traits to improve authentication security and accuracy, since they can be gathered concurrently.

In vein image analysis preprocessing is indispensable. As displayed in Fig. 6.6, the acquired image contains not only the vein pattern but also other tissue, shadows, and noise. Therefore, different approaches for vein pattern extraction were proposed, which most frequently extract the vein pattern as a binary image, see, for example, [36, 37, 41, 42].

For vein pattern recognition, both global and local features are used as well. Usually the courses of the veins or image moments are used as global features [40, 41]. The intersections and endings of veins can be used as local features and are also called minutiae. Features of veins are very similar to features of fingerprints but further detailed features are not suggested, especially because this information gets lost during vein pattern extraction. Also, the adaptation of methods originally used for feature extraction of fingerprints is possible, as shown in [43]. The number of minutiae depends highly on the body part applied for acquisition. As shown in [44], the average number of minutiae found in vein patterns from the back of hand is about 50–70 and in wrist vein patterns approximately 150. This suggests that the number of minutiae in finger vein images is even less than in fingerprints and may not be sufficient in some cases.

Image-based methods are used in vein recognition as well [33]. In order to apply image-based methods, suitable preprocessing methods are needed because a lot of textural information gets lost during binarization. However, this information is essential for image based descriptors.

As the features are similar between fingerprints and veins, the desired properties are similar too. In addition to the two-dimensional translation and rotation, another serious transformation applies through twists of the finger along the longitudinal axis. This transformation also happens during fingerprint acquisition but it is not so critical since fingerprints are to some extend two-dimensional patterns. For vein patterns, the rotation along the longitudinal axis changes the three-dimensional (3D) perspectives from which the patterns are acquired. Hence, the pattern changes entirely and veins could be covered by each other. Because of that, the features should either be robust against this distortion or the sensor construction should prevent excessive transformations of this kind.

6.3 Feature Extraction in Fingerprint Biometrics

6.3.1 Global Features

Global features have a lack in locality, which leads to disadvantages as already discussed in Sect. 6.2.2.2. Hence, image subdividing is applied to improve this deficiency. Naive image subdividing, or in other words dividing the image without respect to the occurred transformations, will harm the invariance properties. Even invariant features like the seven moments from Eq. (6.7) will be dissimilar, since the features are constructed from regions with different content. In order to achieve a translation invariant image subdividing, the sampling grid could be aligned to a center point as illustrated in Fig. 6.7. The additional scale normalization and rotation could guarantee even more robustness against affine transformation. The usage of invariant subdividing resembles local feature extraction, but the subimages contain not meaningful information also. Compared to that, local features are constructed directly from keypoints and regions, so that they represent the significant image parts. Admittedly, global features can also be used as local feature descriptors.

Local techniques for feature extraction used in fingerprint or vein biometrics can be separated into *minutiae-based* and *image-based* approaches, based on the kind of features utilized. In the following subsections these approaches are described.

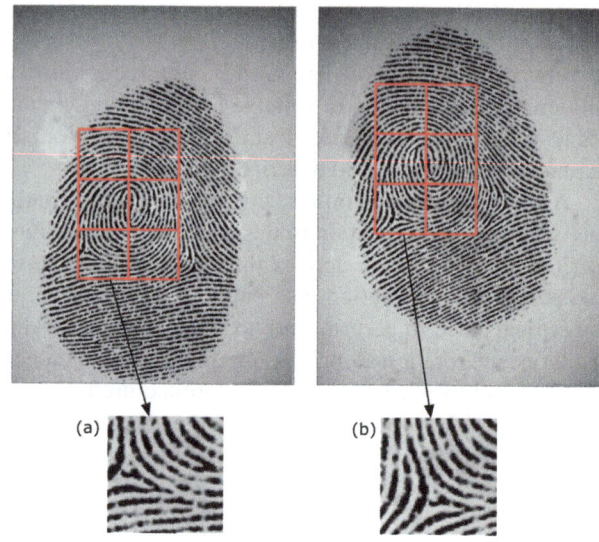

Fig. 6.7 Naive subdividing of two fingerprint images using a 2 × 3 quadratic grid centered on the fingerprint core point as an example. It can be seen that the content of *square* (**a**) and (**b**) is not equal

6.3.2 Minutiae Based Feature Extraction

6.3.2.1 General

Minutiae-based approaches are commonly used for feature extraction in fingerprint and vein biometrics. Minutiae-based methods need to find minutiae points initially, and afterward a feature vector is created based on these points and additional information of the minutiae neighborhood. Also, the term *minutiae representation* is known as a synonym for minutiae descriptor.

As a difference to image based methods, it is assumed that a minutiae template (e.g., according to ISO/IEC 19794-2 [24] or ANSI/NIST-ITL-1 [30]) is already extracted and can be used for the determination of descriptors. The usage of minutiae templates limits the descriptor information to minutiae, which means that no additional information from the image, like gray values, are considered during the descriptor generation. Therefore, minutiae descriptors focus on topological information of the minutiae to describe uniquely the underlying pattern.

6.3.2.2 Minutiae Detection

A minutia, $m_i = \{x_i, y_i, \theta\}$, is defined by its coordinates (x_i, y_i) and the orientation θ specified by the tangent angle with the horizontal axis as shown in Fig. 6.8. According to [21] methods for automatic minutiae detection from fingerprints can be classified into two groups: methods that extract minutiae using binarization and direct gray value based approaches. A literature survey of different minutiae detection methods

Fig. 6.8 A termination minutia with coordinates (x_0, y_0) and tangent angle θ. (Referring to [46])

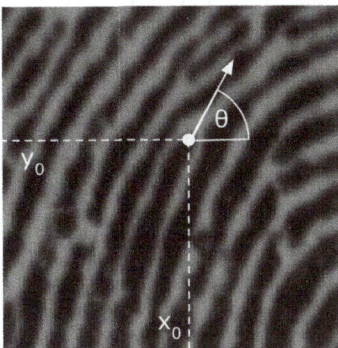

for fingerprint biometrics can be found in [45]. In the following paragraphs the focus is on fingerprint minutiae detection, since the methods used for vein minutiae detection are very similar [39].

Binarization Based Methods At first, these methods convert the recorded gray value image into a binary image. A threshold is used to determine whether a pixel is set to black or white. Some approaches apply a thinning procedure on the binary image, which generates a skeleton with only one pixel line width. Thus, minutiae can be detected by observation of each pixel and the surrounding neighborhood of pixels. The procedure is shown in Fig. 6.9.

A disadvantage of this approach is the loss of information during binarization and thinning. Furthermore, a lot of spurious minutiae are detected by this method due to binarization, especially in the presence of noise. This can also be seen in Fig. 6.9.

Direct Gray Value Based Methods Maio and Maltoni [46] suggested direct gray value minutiae detection because of these drawbacks. Here, the idea is to follow the ridge lines directly on the gray value image. A set of starting points is determined, and the endings and bifurcations of the ridge lines are detected by ridge line following using the directional image. An additional labeling algorithm is needed to prevent the algorithm from considering lines twice. Some modifications of this algorithm can be found in [21]. These methods are very application oriented, since they make use

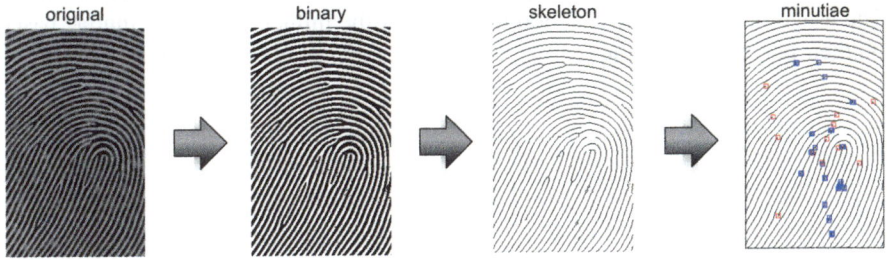

Fig. 6.9 Minutiae detection based on binarization. (Referring to [46], [47])

of the knowledge about the underlying pattern. Ridge line following is not the only way of direct gray value minutiae detection. Other methods use window functions and try to determine whether the windows content is a minutia or not, for example, by fuzzy or neural network based classifiers [48].

These methods are more related to general image based feature detection approaches as they will be described in Sect. 6.3.3.1.

6.3.2.3 Minutiae Description

The next step after minutiae detection, or with a provided minutiae template, is minutiae description using topological or semantic relations between minutiae points. A lot of methods were proposed for local minutiae description. Maltoni et al. [21] distinguish between *nearest neighbor-based* and *fixed radius-based* minutiae descriptors.

Nearest Neighbor-Based Minutiae Descriptors This kind of descriptor characterizes the minutiae environment by help of k nearest neighbor minutiae. In order to obtain invariant feature vectors, attributes like the distance between minutiae, angle of connection lines or the number of ridges between the minutiae are used. Here, it should be noticed that, for example, distances are not invariant under scale transformations. This can be neglected in fingerprint biometrics due to the fact that scale changes are not common, if sensors of the same resolution are used.

The order of the neighbor minutiae is a critical point of this kind of descriptor. Local image distortions may lead to different feature vectors and therefore to classification errors, if the order is based on the distances from the neighbors to the central minutia and these distances are small or the same for two neighbors.

Fixed Radius-Based Minutiae Descriptors A neighborhood can be obtained by the use of a fixed radius r around a central minutia, which includes all minutiae whose distance to the central minutia is smaller than r. Subsequently, this neighborhood can be described using invariant attributes of the minutiae relations. A problem of this method is the treatment of minutiae on the regions edge. This minutiae could be inside the region in the first sample and outside in the next one due to local distortion or noise, which leads to defective feature vectors. This problem can be counteracted by multiple analyses with slightly enlarged radius, but the basic problem remains. The usage of a weighting function that weights the influence of the minutiae based on their position inside the region could be another possible solution.

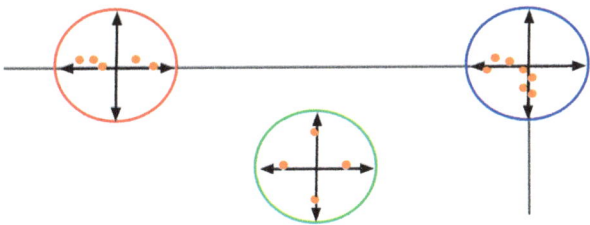

Fig. 6.10 The idea of the Harris corner detector. On edges the image signal changes in one dimension (*left circle*), at corners it changes two dimensional (*right circle*). The image signal does not change on uniform areas (*centered circle*)

6.3.3 Image Based Feature Extraction

6.3.3.1 Image Based Feature Detection

A feature detector finds interesting points or regions in images. A keypoint is defined as a point where the image signal changes intenty, like corners, T-junctions, Y-junctions and much more [16, 49]. These regions need to be determined before description, since local features are defined as a tuple of a keypoint or region and a characteristic description of the neighborhood (see Sect. 6.2.2.2).

The same keypoints need to be detected independently in two images and the determined regions should be invariant with respect to the occurring transformations in order to get alignment-free features. Because of that, these regions are often described by circles, or in case of general perspective transformations by ellipses or parallelograms.

Image-based feature detectors can be divided into three classes by the kind of keypoints or regions they detect.

Corner Detectors Keypoint detectors are most likely named corner detectors due to the definition of keypoints. Corner detectors are explored extensively, and a lot of different approaches were proposed. The Harris corner detector proposed by Harris and Stephens [50] is probably the most prominent detector of this kind. Other examples are Features from accelerated segment test (FAST) [51], adaptive and generic accelerated segment test (AGAST) [52], or Oriented FAST and Rotated BRIEF (ORB) [53]. A detailed review of different methods can be found in [49, 54].

Take, for example, the Harris detector to get a better understanding about corner detection. The idea of this detector is to examine every pixel within the image to find out if the image signal in a small window around this pixel is changing in both directions, as it is displayed in Fig. 6.10.

The *second moment matrix* is used to determine the properties of the signal inside of the pixel's window. This is possible, since the central image moments (see Eq. (6.5)) can be used to determine the direction of the strongest respective weakest signal changes [9]. More precisely, the eigenvalues of the second moment matrix are used. Thus, the keypoints obtained are invariant with respect to translation and rotation, so that this transformation do not affect the detection algorithm.

Fig. 6.11 Example of the SIFT-detector used on a fingerprint image. The different size and direction of the circles result from the detected extrema in scale space

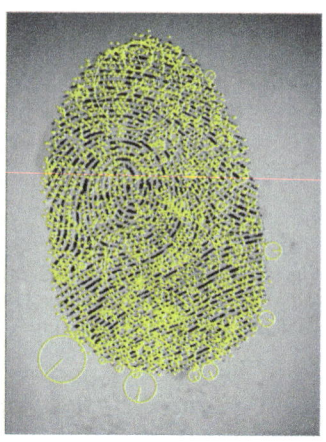

Regions for feature description are required in order to use this keypoints as local features. For example, circles with a fixed size centered on the detected keypoints can be utilized. The regions detected this way are not scale invariant, since the region size is not related to the image scale. Mikolajczyk and Schmid [18] proposed a scale invariant corner detector based on the Harris detector, which is called *Harris-Laplace detector*.

Blob Detectors A blob detector, in contrast to corner detectors, tries to identify blob like structures with a high amount of information directly. The most prominent blob detector was proposed by Lowe [55, 56] and is called *scale-invariant feature transform (SIFT)*. An example is shown in Fig. 6.11.

The SIFT detector identifies interesting blobs by determination of scale space extrema using the *Difference of Gaussians (DoG)* operator. The scale space theory is broadly described in [57]. In scale space, the image is presented in different scales (using Gaussian filter), which makes the estimation of the DoG very simple, as illustrated in Fig. 6.12. Regions extracted by the SIFT detector are invariant under changes of translation, rotation and scale.

Some blob detectors with the same properties of invariance are for instance the *Hessian-Laplace* detector [18] or the *speeded up robust features*-detector *(SURF)* [58]. Other blob detectors are also invariant under general affine transformations like the Harris- and Hessian-affine detectors proposed in [18].

Region Detectors Some feature detectors are able to extract affine invariant regions of different forms like the *maximally stable extremal regions* (*MSER*) proposed by Matas et al. [59]. This kind of detectors was not used for fingerprint or vein biometrics, since the locality and distinctiveness of these regions may not be suitable for these particular patterns and therefore they are not considered further. A wide overview of different feature detectors can be found in [17].

Fig. 6.12 Illustration of the difference of Gaussian's operator

6.3.3.2 Image-Based Feature Description

The image-based feature descriptors are constructed from the available image information inside of the detected regions. The usage of these intensively researched techniques is obvious, since the biometric traits are acquired as images. As already mentioned in Sect. 6.3.1, the image moments can be used as local image based descriptors. In addition, a wide range of image based descriptors exists and a couple of them are based on the ideas of Lowe's [55, 56] SIFT descriptor (see for instance [19, 60]). The SIFT descriptor uses gradient histograms, calculated from an interesting points neighborhood with respect to rotation and scale. The gradient histogram represents the frequency distribution of the gradient of an image section in relation to its magnitude, where the gradient describes the direction of the strongest gray value amendment and its magnitude represents the amendment's intensity.

It is to investigate whether due to the characteristics of these patterns image based descriptors provide satisfying results. Perhaps description based on the relations of keypoints, like minutiae description, is more suitable in this context than using the image-based information.

6.4 A Comparison of Alignment-Free Biometric Systems

6.4.1 Global Methods

Some difficult problems and interesting improvements are derived from the usage of global features, even though global methods cannot be considered as state of the art in fingerprint and vein biometrics.

Vein Pattern Recognition Using Invariant Moments
A feature extraction method for vein pattern recognition based on invariant moments is described in the work of Xueyan et al. [41]. At first, the image is preprocessed using *wavelet transforms*, and then the vein pattern is extracted. Subsequently, Xueyan et al. [41] utilized Hu's [12] invariant moments and additional five invariant moments derived from [61] as features. Based on their experiments, Xueyan et al. [41] decided not to use Hu's moments [12] of order 5 and 7, since they were too different. Because

of that, the authors rely only on the others moments, which leads to a ten-dimensional feature vector.

However, this raises the question whether a ten-dimensional feature vector is sufficient for classifications of large databases. Depending on the used classifier, a specific feature vector dimension ratio must be maintained in order to achieve satisfactory classification results [3]. Therefore, the image subdividing can be used to improve the number of features.

Pseudo-Zernike Moments and Image-Subdividing
Deepika et al. [62] suggest a fingerprint recognition system based on pseudo-Zernike moments. First, the region of interest is extracted, and the pseudo-Zernike moments of this region from order 1–5 are calculated. The image is subdivided to increase the quantity of features. The pseudo-Zernike moments of order 5 perform the best for the whole image, whereas order 3 moments with 10×10 sub-images perform even better, according to their experiments.

However, Deepika et al. [62] do not suggest how to determine the sub-images. If simple image partitioning is used, it will lead to serious drawbacks in regards of feature invariance as mentioned at the beginning of Sect. 6.3.1. In order to obtain alignment-free features, the analyzed sub-images must be stated invariant.

Invariant Moments and Invariant Subdividing
Yang and Park [63] describe a method for fingerprint analysis, which uses invariant moments and invariant image subdividing. At first, the image is preprocessed using the *short-time Fourier transform* according to [64]. A reference point is extracted by the use of complex filters and its location is determined by the application of *least mean square orientation estimation* (LMS) adapted from [65]. The reference point location is subsequently used to achieve an invariant image subdividing. Yang and Park [63] divide the fingerprint recording around the reference point into four squares and thus receive 28 invariant features by using the seven Hu-moments [12].

Looking at the methods presented so far, it is clear that the drawbacks of global features can only be counteracted by appropriate auxiliary techniques.

6.4.2 Image Based Methods

Fingerprint Analysis with Alignment-Free Local SIFT-Features
Park et al. [32] were the first who have used Lowe's SIFT algorithm [55, 56] for fingerprint analysis. They aimed at good results in fingerprint recognition too, since SIFT has proven itself in many applications for object recognition. The images are preprocessed by adjustment of the gray-level distribution before interesting regions are identified using the SIFT detector, as it was discussed in Sect. 6.3.3.1. An advantage of the SIFT detector is that it finds more keypoints than a minutiae detector [35] and these keypoints include also minutiae and pores [34].

The corresponding SIFT descriptor is computed for each detected SIFT keypoint. Park et al. [32] used a point wise comparison of the SIFT keypoints based on the descriptors using the Euclidean distance. False matches that arise from the comparison must be removed. The SIFT algorithm performs worse in comparison to a not further specified minutiae based approach on the FVC 2002 (*Fingerprint Verification Competition*) DB 1 and 2 [66], with an equal error rate of 8.44 % respectively 10.76 %, against 1.79 % and 2.13 %. That is why Park et al. [32] proposed a fusion of both algorithms according to [67], that achieves an equal error rate of 0.99 % and 1.07 %. In addition, they assume that the results can be improved by appropriate preprocessing and matching techniques.

Fingerprint Indexing with Reduced SIFT-Features
Shuai et al. [68] used the SIFT algorithm for fingerprint indexing. The number of features are reduced by appropriate thresholds and only the N most significant features are used for indexing. The index is computed using *locality-sensitive hashing* (LSH) according to [69].

Finger Vein Analysis Using SIFT-Features
Pang et al. [33] suggested SIFT for the analysis of finger vein patterns. Since preprocessing is very important in vein recognition (see Sect. 6.2.3.4), the vein pattern has to be extracted before further processing. At first, the background is removed, and after that the contrast between the veins and the rest of the image is increased using a *histogram equalization*. The features of the vein pattern can be extracted using SIFT after preprocessing.

The aim of Pang et al. [33] was the extraction of rotation invariant features that makes a biometric system based on finger veins particularly user friendly. They have tested the proposed system on a database, developed by themselves, including 95 individuals with 11 recordings of each individual. Also they compared their system with the not rotation invariant *local line binary pattern* (LBP) proposed by [70]. The results by Pang et al. [33] show that their system performs better under rotation then the LBP. However, an evaluation on a common database is needed in order to obtain an adequate comparison.

Other Local Feature Extraction Methods
He et al. [34] demonstrated, by the example of *SURF*, that not only SIFT is generally suitable for the usage in biometric systems. Different image-based feature detectors, like *Harris* [50], *Hessian* [18], *FAST* [51, 71], *SURF* [58] and many more, can be applied in order to obtain invariant feature points. These methods can increase the feature vectors distinctiveness, especially for biometric traits with a little amount of minutiae, since these methods detect more keypoints then minutiae detectors. It is to investigate, which detectors respectively which keypoints are the most suitable for a specific biometric system as minutiae are known to have a high information content. The suitability depends on the required invariance, as well as the keypoint type and the corresponding region information.

The variety of proposed invariant image based feature descriptors (e.g., *moments* [9, 10, 12, 15], *SIFT* [55, 56], *SURF* [58], *Gradient Location and Orientation Histogram (GLOH)* [19], *Histogram of Oriented Gradients (HoG)* [60]) offers possibilities for further research. The example [68] shows that the required properties of a descriptor differ depending on the application.

6.4.3 Minutiae-Based Methods

As mentioned in Sect. 6.3.2.1, most proposals of minutiae based biometric systems assume that a minutiae template (e.g., ANSI/NIST-ITL-1 [30] or ISO/IEC 19794-2 [24]) is already extracted and can be used for minutiae descriptor generation. Therefore, this section describes systems which apply alignment-free minutiae based feature description.

Minutiae Description by Local Coordinate Systems
Chikkerur et al. [72] propose a translation and rotation invariant minutiae descriptor, called *K-plet*. It is a nearest neighbor descriptor, which uses a central minutiae $m_i = (x_i, y_i, \theta_i)$ (according to ISO/IEC 19794-2 [24]) and K neighboring minutiae $\{m_1, m_2, \ldots, m_k\}$. The authors suggest two possibilities of how the neighborhood is chosen. According to their first suggestion, simply the K nearest neighbors are chosen, which leads to the drawbacks mentioned in Sect. 6.3.2.3. In addition, they suggested a sequential selection of the neighborhood of the four quadrants around the central minutiae. This methods should guarantee that a wide range of the fingerprint is considered, and the global structure is also taken into account.

Each neighbor minutiae is defined by $m_j = (\phi_{ij}, \theta_{ij}, r_{ij})$ where $j = 1, 2, \ldots K$. Where r_{ij} represents the Euclidean distance between the minutia m_j and the central minutia m_i, which can be seen as a edge connecting the minutiae with a length of r_{ij}. ϕ_{ij} describes the orientation of this edge and θ_{ij} is defined as the relative orientation of m_j with respect to m_i. The angles are measured as the orientation of m_i representing the x-axis.

An advantage of the K-plet representation is that it can be easily transformed into a graph. This representation allows a global consolidation by an extended *breadth first search*, which is referred as *Coupled BFS* by the authors. The proposed system was evaluated by the authors using the FVC 2002 DB1 [66] and they achieved an equal error rate of 1.5 %. However the used parameter K or an advice for a suitable value is not mentioned.

Minutiae-Description Based on Spatial and Directional Contributions
Another minutiae descriptor was proposed by Cappelli et al. [73]. They describe a very detailed fixed-radius descriptor, which characterizes the local structure by different spatial and directional contributions of the minutiae. Each minutia is represented by a *Cylinder*, i.e., a 3D data-structure, which is generated on the basis of invariant angles and distances of their neighborhood minutiae. For this reason, the descriptor is called *Minutia Cylinder-Code* (MCC). The prerequisite is a minutiae template according to ISO/IEC 19794-2 [24]. The cylinder can be illustrated as horizontal slices, which reflect the spatial and directional contributions of the neighborhood minutiae with respect to the orientation of a central minutia. The considered orientation is changed as the cylinder grows vertical, which results in other spatial and directional contributions. Together, all horizontal slices with different orientations form the cylinder and consequently the minutia descriptor. It is referred to [35] or [73] for more information about the computation of cylinder sets.

The feature vectors generated using this method can be compared with simple metrics. Cappelli et al. [73] also suggest different possibilities for the computation of a global score for classification. A detailed evaluation of this method applied on the FVC 2006 DB2 [74] can be found in the original paper. Also, different unspecified algorithms used for minutiae detection and the proposed matching techniques are evaluated by the authors. The MCC is a highly parameterized descriptor, but Cappelli et al. [73] proposed a list of parameter assignments, which they improved in [75]. In addition, this descriptor can be used for template-protection [76], indexing [77] and it is adaptable for chip cards.

Analysis of Vein Patterns Using Local Minutiae and Mellin Transform
The method proposed by Hartung et al. [39, 44] uses local minutiae in vein patterns to generate translation, rotation, and scale invariant features. At first, the contrast is increased by *adaptive non-local means,* and afterwards a *nonlinear diffusion algorithm* is applied to improve the structure and decrease the noise. A *multi-scale filter* is used to separate the vein pattern from the background. At next, the image is binarized, and the minutiae are extracted, like in the binarization based approaches for fingerprint minutiae detection mentioned in Sect. 6.3.2.2. The *spectral minutiae representation* is used as descriptor, which is based on the *Mellin transform*. The invariance properties can be achieved easier due to the transformation to the frequency domain. Spectral minutiae representation was originally proposed by Xu et al. [78, 79] to be applied as fingerprint minutiae descriptor. This example shows how similar the approaches are in fingerprint and vein biometrics.

Vein Minutia Cylinder-Code
Hartung et al. [43] also adapted the MCC [73] for the usage in vein recognition systems. Their paper describes the selection of appropriate parameters for a *Vein Minutia Cylinder-Code* (VMCC). In the original paper of Cappelli et al. [73] there is a lack of explanation how and why the parameters for the MCC are chosen, even though they have specified the parameters. Hartung et al. [43] assume that the originally proposed parameter set is not optimal for vein patterns due to the trait related differences. Moreover, they hypothesize that even different parameters provide optimal results for vein patterns from different body regions.

A *genetic algorithm* was used to optimize the 26 parameters from the original method. This kind of algorithm is inspired by the natural evolution and can be used as optimization method for hard problems and unknown coherence, if no iterative solution is known. Overall, the authors have trained their system on the three databases: *SNIR*(back of the hand veins), *SFIR*(back of the hand veins), *UC3M*(wrist veins). They reached the conclusion that various parameter sets are optimal in the case of different body regions. In addition, the parameters determined by the genetic algorithm provide better classification results ($EER_{SNIR} = 1,45\%$, $EER_{SFIR} = 1,88\%$, $EER_{UC3M} = 0,31\%$) for vein patterns than the original proposed ones ($EER_{SNIR} = 2,06\%$, $EER_{SFIR} = 3,15\%$, $EER_{UC3M} = 0,31\%$). The optimized parameter set can be found in [43].

The papers of Hartung et al. [43, 44] show clearly that minutiae based feature extraction methods for fingerprint recognition can be successfully adapted for vein pattern analysis.

6.4.4 Summary

The feature extraction methods presented in this Section are summarized in the following Table 6.1, which allows a comparison of the methods based on: Feature types, biometric traits and the used preprocessing techniques as well as the experimental error rates stated by the authors and the corresponding databases.

6.5 An Evaluation Strategy for Local Features

6.5.1 Evaluation of Local Feature Extractors

Each kind of local feature has different advantages and disadvantages, as already discussed. Appropriate criteria are needed to compare the different feature extraction methods for a particular application. In general, the system error rates are used for comparison if one attempts to choose a feature extraction method. However, the error rates are rating the entire system and they are not an explicit criterion for the quality of the used feature extractor. Biometric systems consist of different stages and the error rates are influenced by each phase as described in Sect. 6.2. Thus, the error rates can be used as a suitable evaluation of the quality and interaction of all system components. An evaluation of the particular detector and descriptor is necessary in order to analyze which feature extraction method is generally suitable for the use in biometric systems or how already introduced methods can be improved.

Biometric systems are essentially pattern recognition systems, and the different approaches like image based and minutiae based methods are similar to some extent, although they arose from different disciplines. Therefore it is reasonable to adapt an evaluation strategy for local feature extraction from the field of digital image processing.

6.5.2 Repeatability

The important properties of reliable local features have been described in Sect. 6.2.2.2. Given two images of the same object, a feature detector should detect a high amount of features that are visible in both images, even under changes of the objects position or in presence of noise. This property is called *repeatability*, and can be used as a criterion for the robustness and invariance of a feature detector. At this point, it makes no difference whether minutiae or other keypoints are detected. However, repeatability is not a criterion of the information quality with respect to the classification results.

The repeatability criterion was originally proposed by Schmid et al. [49] and is defined as follows (according to [49]):

Table 6.1 A summary of the biometric systems considered in Sect. 6.4

Source	Feature type	Trait	Method	Pre-processing	Stated evaluation	Used database
Deepika et al. (2010)	Global	Fingerprint	Pseudo-Zernike moments (order 5)	ROI-extraction, image subdividing	9.7 % EER 0.903 true positive (TP) 0.008 False Positive (FP)	FVC 2002 DB1 Set A [66]
Xueyan et al. (2007)	Global	Fingervein	Hu-Moments, generalized moments (Tian-Xu 2004)	Normalization, dyadic wavelet transform, denoising by soft-thresholding	7.3 % EER	Custom: 256 images, from 32 persons, index and middle finger
Yang and Park (2008b)	Global	Fingerprint	Invariant subdividing (2×2), 7 Hu moments	Short-time fourier transform, enhancement	EER = 3.69 %	Complete FVC 2002 [66]
Park et al. (2008)	Local	Fingerprint	SIFT, trimming false matches	Adjusting the gray level distribution	EER = 8.44 % (DB1), EER = 10.76 % (DB2)	FVC 2002 DB 1, 2 [66]
Park et al. (2008)	Local	Fingerprint	Fusion: SIFT, Minutiae	Adjusting the graylevel distribution	EER = 0.99 % (DB1), EER = 1.07 % (DB2)	FVC 2002 DB 1, 2 [66]
Pang et al. (2012)	Local	Fingervein	SIFT	Segmentation and histogram equalization	EER = 1.71 %	Custom: 95 classes, with 11 finger vein images each
Chikkerur et al. (2005)	Local	Fingerprint	ISO-template, Kplet-descriptor, Coupled breadth first search (CBFS)	Not specified	EER = 1.5 %	FVC 2002 DB1 [66]

Table 6.2 (continued)

Source	Feature type	Trait	Method	Pre-processing	Stated evaluation	Used database
Cappelli et al. (2010a)	Local	Fingerprint	ISO-template, minutia cylinder-code (MCC16)	Not specified	2.66 % EER Local Similarity Sort (LSS), 2.32 % EER Local Similarity Assignment (LSA), 1.32 % EER Local Similarity Sort with Relaxation (LSS-R), 1.18 % EER Local Similarity Assignment with Relaxation (LSA-R)	FVC 2006 [74]
Cappelli et al. (2010b)	Local	Fingerprint	ISO-template, minutia cylinder-code (optimized)	Not specified	Lowest EER = 0.121 %	FVC2006 DB2 [74]
Hartung et al. (2011)	Local	Veinpatterns	Minutiae-detection (binary image), SML	Adaptive nonlocal means, nonlinear diffusion algorithm, multi-scale filtering	EER = 1.62 %, 4.33 %, 5.90 %	Singapore NIR (SNIR), Singapore FIR (SFIR), University Carlos III of Madrid (UC3M) [39]
Hartung et al. (2013)	Local	Veinpatterns	Minutiae-detection (binary image), VMCC	Adaptive nonlocal means, nonlinear diffusion algorithm, multi-scale filtering	EER = 1.45 %, 1.88 %, 0.31 %	SNIR, SFIR, UC3M [39]

EER equal error rate, *FVC* fingerprint verification competition, *ISO* International Organization for Standardization, *SIFT* scale-invariant feature transform, *ROI* region of interest, *VMCC* vein minutia cylinder-code

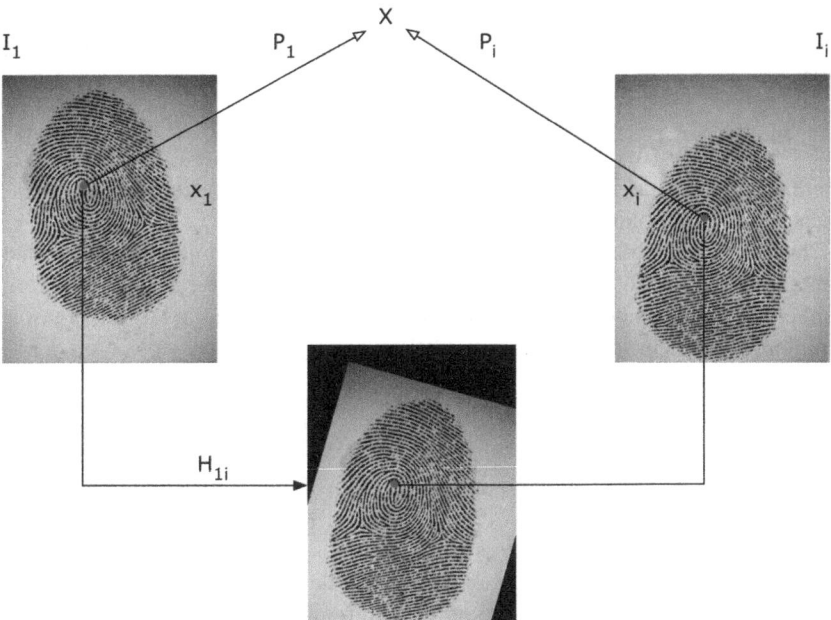

Fig. 6.13 Illustration of the homography between to fingerprint images. (With reference to [49])

Let X be a 3D point and P_1, P_i two projection matrices. The projection of X onto the images I_1 and I_i is defined as $x_1 = P_1 X$ respectively $x_i = P_i X$. If one can find a *corresponding point* x_i for a given point x_1, the point x_1 is called repeated in I_i. In case of planar 3D scenes, this can be determined by a unique relation between x_1 and x_i:

$$x_i = H_{1i} x_1 \text{ where } H_{1i} = P_i P_1^{-1}. \tag{6.9}$$

The homogeneous 3×3 Matrix H is called *homography*. For illustration see Fig. 6.13.

As shown in Fig. 6.13, the homography is used to perform an image alignment, which brings the problem of affine invariant pattern recognition back to image alignment. An accurate alignment is possible, if the homography is known exactly, for example, due to a special experimental construction, which results in a precise evaluation.

The repeatability criterion should only consider the points that are actually present in both images. In order to achieve this, a common image area is determined with the help of the homography (according to [49]):

$$\{\tilde{x}_1\} = \{x_1 \mid H_{1i} x_1 \in I_i\} \text{ and } \{\tilde{x}_i\} = \{x_i \mid H_{i1} x_i \in I_1\}, \tag{6.10}$$

where $\{x_1\}, \{x_i\}$ are the detected points in I_1, I_i. H_{ij} is defined as the homography between I_i, I_j and $\{\tilde{x}_1\}, \{\tilde{x}_i\}$ are the points inside of the common image area.

A neighborhood property is applied, because the point positions are probably not exactly the same. A pair of corresponding points $(\tilde{x}_1, \tilde{x}_i)$ is defined by (according to [49]):

$$R_i(\epsilon) = \{(\tilde{x}_1, \tilde{x}_i) \mid dist(H_{1i}\tilde{x}_1, \tilde{x}_i) < \epsilon\}, \tag{6.11}$$

where ϵ is defined as the relative point position error. According to [18], $\epsilon = 1,5$ pixels should be applied.

Thus, the repeatability score $r_i(\epsilon)$ can be defined as the number of corresponding points with respect to all points inside of the corresponding region. However, the number of points can be different from one image to the other and for this reason the minimum value is used (according to [49]):

$$r_i(\epsilon) = \frac{|R_i(\epsilon)|}{min(n_1, n_i)}, \tag{6.12}$$

where $n_1 = |\{\tilde{x}_1\}|$ and $n_i = |\{\tilde{x}_i\}|$ are the number of points inside the common area of the images I_1 and I_i.

So far the criterion just considers point positions, which is only suitable for interesting point detectors (e.g., minutiae, corners). Mikolajczyk and Schmid [18] extend this criterion for the usage with general affine region detectors, by considering the corresponding regions overlap.

A pair of corresponding points $(\tilde{x}_1, \tilde{x}_i)$ is now defined as (according to [18]):

$$R_i(\epsilon, \epsilon_s) = \{(\tilde{x}_1, \tilde{x}_i) \mid dist(H_{1i}\tilde{x}_1, \tilde{x}_i) < \epsilon \wedge \epsilon_s < 0,4\} \tag{6.13}$$

The overlap error ϵ_s is defined as (according to [18]):

$$\epsilon_s = 1 - \frac{\mu_1 \cap (A^T \mu_i A)}{(\mu_1 \cup A^T \mu_i A)}, \tag{6.14}$$

where μ_1, μ_i are the elliptic regions around the points $\tilde{x}_1, \tilde{x}_i$. The locally linearized homography in point $\tilde{x}_i$ is defined as A, and $\mu_1 \cap (A^T \mu_i A)$ is the intersection of the two regions, respectively $(\mu_1 \cup A^T \mu_i A)$ is the union.

6.5.3 1-Precision-Recall

Another important property of local features is the information content. Features should make patterns distinguishable, but this can only be achieved if the pattern is described uniquely. However, information content is not easy measurable compared to repeatability. On the one hand, the detector affects the regions and on the other hand the information must be suitably represented by the descriptor for further processing by the classifier.

Mikolajczyk and Schmid [19] suggested the measures *1-precision* and *recall* for the comparison of different descriptors. *Precision* and *recall* are used in general for the evaluation of classifiers, therefore $1 - precision$ and $recall$ can be used for distinction between classifier and feature descriptor evaluation. These criteria are based on the number of correct and false matched features during the classification and they are defined as follows (according to [19]):

$$1 - precision = \frac{\#false\,matches}{\#all\,matches} \quad (6.15)$$

$$recall = \frac{\#correct\,matches}{\#correspondences} \quad (6.16)$$

Two regions are considered as match if the distance between their descriptors is under a threshold t. The number of correct respectively false matches can be determined with help of the overlap error (Eq. 6.14), where $\epsilon_s < 0,5$ should apply. The number of corresponding points (i.e., $\#correspondences$) can be determined like for the repeatability criterion. One can achieve $1 - precision, recall$ curves through variation of the classification threshold t. These curves can further be used to compare different feature extraction methods.

The measurements $1 - precision, recall$ represent the quality of the used feature descriptor as well as feature detector and classifier. The interaction of all these components can be assessed by the use of this evaluation technique. If the feature extraction methods are changed, but the same classifier is used, a feature based evaluation is achieved, and the methods can be compared directly.

Different local feature extraction methods in biometrics can be compared explicitly with the help of this evaluation strategy. It can be applied on every kind of local feature detector and -descriptor, although it was proposed for image based feature extraction. Hence, it helps to find appropriate methods for a given application.

6.6 Conclusion

In this chapter methods for alignment-free feature extraction in finger and vein biometrics were presented. The basics of image-based pattern recognition and biometrics were combined to create a common understanding of features and feature extraction. Therefore, the attention was drawn to the similarities and differences of the various approaches. All techniques have advantages and disadvantages. Minutiae-based approaches are sensitive to images with a low amount of minutiae, for example, due to noise or scares, although minutiae based approaches are the most researched ones, and the proposed systems show good error rates. However, minutiae based descriptors are complex and designed to describe the available information by minutiae relations. Against minutiae-based approaches, the number of keypoints detected by image-based feature detectors is higher, but it is to investigate, whether image-based descriptors are suitable to describe the biometric trait uniquely.

The proposed evaluation strategy can be applied to compare and assess different feature extraction methods and especially their robustness and invariance in detail, which will help to find a suitable method for a particular application. In addition, the used sub-algorithms, for example, different preprocessing techniques, and their influence on the feature extraction can also be examined to further optimize the overall system.

In addition, hybrid approaches are feasible due to the advantages and disadvantages of minutiae- and image-based methods , respectively. Thus, more feature points can be obtained by image-based detection, which can then be described by minutiae-based descriptors. These approaches can also be evaluated using the proposed criteria.

References

1. Wood J. Invariant pattern recognition: a review. Pattern Recognit. 1996;29(1):1–17.
2. Steger C, Ulrich M, Weidemann C. Machine vision algorithms and applications. Weinheim: Wiley-VCH 2008.
3. Theodoridis S, Koutroumbas K. Pattern recognition. 4th ed. Amsterdam: Elsevier Academic Press; 2008.
4. Guyon I, Elisseeff A. An introduction to variable and feature selection. J Mach Learn Res. 2003;3:1157–82.
5. Nixon M, Aguado AS. Feature extraction & image processing. Amsterdam: Elsevier Academic Press; 2008.
6. Bishop CM. Neural networks for pattern recognition. Oxford University Press; 1995.
7. Cortes C, Vapnik V. Support-vector networks. Mach Learn. 1995;20(3):273–97.
8. Bishop CM, et al. Pattern recognition and machine learning. vol. 1. New York: Springer; 2006.
9. Prokop RJ, Reeves AP. A survey of moment-based techniques for unoccluded object representation and recognition. CVGIP: Graph Models Image Process. 1992;54(5):438–60.
10. Teague MR. Image analysis via the general theory of moments*. JOSA. 1980;70(8):920–30.
11. Mercimek M, Gulez K, Mumcu TV. Real object recognition using moment invariants. Sadhana. 2005;30(6):765–75.
12. Hu MK. Visual pattern recognition by moment invariants. IRE Trans Inf Theory. 1962;8(2):179–87.
13. Yang J. Biometrics. Non-minutiae based fingerprint descriptor, chapter 4. InTech, June 2011.
14. Teh C-H, Chin RT. On image analysis by the methods of moments. IEEE Trans Pattern Anal Mach Intell. 1988;10(4):496–513.
15. Khotanzad A, Hong YH. Invariant image recognition by zernike moments. IEEE Trans Pattern Anal Mach Intell. 1990;12(5):489–97.
16. Schmid C, Mohr R. Local grayvalue invariants for image retrieval. IEEE Trans Pattern Anal Mach Intell. 1997;19(5):530–5.
17. Tuytelaars T, Mikolajczyk K. Local invariant feature detectors: a survey. Found Trend® Comput Graph Vis. 2008;3(3):177–280.
18. Mikolajczyk K, Schmid C. Scale & affine invariant interest point detectors. Int J Comput Vis. 2004;60(1):63–86.
19. Mikolajczyk K, Schmid C. A performance evaluation of local descriptors. IEEE Trans Pattern Anal Mach Intell. 2005;27(10):1615–30.
20. Jain AK, Ross A, Prabhakar S. An introduction to biometric recognition. IEEE Trans Circuit Syst Video Technol. 2004;14(1):4–20.
21. Maltoni D, Maio D, Jain AK, Prabhakar S. Handbook of fingerprint recognition. London: Springer; 2009.
22. Jain AK, Flynn P, Ross AA. Handbook of biometrics. London: Springer; 2010.

23. Bundesamt für Sicherheit in der Informationstechnik (BSI). BioKeyS Pilot-DB Teil 2 (Projekt Template Protection), Abschlussbericht. 2011. https://www.bsi.bund.de/SharedDocs/Downloads/DE/BSI/Publikationen/Studien/BioKeys/BioKeyS-Abschlussbericht.pdf?__blob=publicationFile. Accessed 15 April 2014.
24. International Organization for Standardization and International Electrotechnical Commission. ISO/IEC 19794.
25. Lee C, Choi JY, Toh KA, Lee S. Alignment-free cancelable fingerprint templates based on local minutiae information. IEEE Trans Syst Man Cybern Part B: Cybern. 2007;37(4):980–92.
26. Rathgeb C, Uhl A. A survey on biometric cryptosystems and cancelable biometrics. EURASIP J Inf Secur. 2011;2011(1):1–25.
27. Juels A, Sudan M. A fuzzy vault scheme. Des Code Cryptogr. 2006;38(2):237–57.
28. Bundesamt für Sicherheit in der Informationstechnik (BSI). Fingerabdruckerkennung. 2013. https://www.bsi.bund.de/SharedDocs/Downloads/DE/BSI/Biometrie/Fingerabdruckerkennung_pdf.pdf?__blob=publicationFile. Accessed 28 Nov 2013.
29. Henry ER. Classification and uses of finger prints. Routledge; 1900.
30. ANSI/NIST. ANSI/NIST-ITL-1-2011. 2011. http://biometrics.nist.gov/cs_links/standard/ansi_2012/Update-Final_Approved_Version.pdf. Accessed 2 Dec 2013.
31. Jain AK, Chen Y, Demirkus M. Pores and ridges: high-resolution fingerprint matching using level 3 features. IEEE Trans Pattern Anal Mach Intell. 2007;29(1):15–27.
32. Park U, Pankanti S, Jain AK. Fingerprint verification using sift features. In: SPIE. 2008; vol. 6944, pp. 69440K.
33. Pang S, Yin Y, Yang G, Li Y. Rotation invariant finger vein recognition. In: Biometric recognition, pp. 151–6. Springer; 2012.
34. He S, Zhang C, Hao P. Comparative study of features for fingerprint indexing. In: 16th IEEE International Conference on Image Processing (ICIP). 2009; pp. 2749–52.
35. Fay R. An analysis of alignment-free feature-extraction methods for fingerprint and vein biometrics. Master's thesis, University of Siegen; 2014.
36. Miura N, Nagasaka A, Miyatake T. Extraction of finger-vein patterns using maximum curvature points in image profiles. IEICE Trans Inf Syst. 2007;90(8):1185–94.
37. Miura N, Nagasaka A, Miyatake T. Feature extraction of finger-vein patterns based on repeated line tracking and its application to personal identification. Mach Vis Appl. 2004;15(4):194–203.
38. Miura BT, et al. Vein extraction methods. 2012. http://www.mathworks.com/matlabcentral/fileexchange/35716-miura-et-al-vein-extraction-methods. Accessed 6 Dec 2013.
39. Hartung D. Vascular pattern recognition: and its application in privacy-preserving biometric online-banking systems. PhD thesis, Gjøvik University College; 2012.
40. Xueyan L, Shuxu G. The fourth biometric-vein, pattern recognition techniques, technology and applications. InTech, 2008.
41. Xueyan L, Shuxu G, Fengli G, Ye L. Vein pattern recognitions by moment invariants. In the 1st International Conference on Bioinformatics and Biomedical Engineering, ICBBE. 2007; pp. 612–5.
42. Hartung D. Venenbilderkennung. Datenschutz und Datensicherheit—DuD. 2009;33(5):275–9.
43. Hartung D, Tistarelli M, Busch C. Vein minutia cylinder-codes (v-mcc). In International Conference on Biometrics (ICB). 2013; pp. 1–7.
44. Hartung D, Olsen MA, Xu H, Busch C. Spectral minutiae for vein pattern recognition. In International Joint Conference on Biometrics (IJCB). 2011; pp. 1–7.
45. Bansal R, Sehgal P, Bedi P. Minutiae extraction from fingerprint images-a review. arXiv preprint arXiv:1201.1422, 2011.
46. Maio D, Maltoni D. Direct gray-scale minutiae detection in fingerprints. IEEE Trans Pattern Anal Mach Intell. 1997;19(1):27–40.
47. Athi. Fingerprint Minutiae Extraction: 2011. http://www.mathworks.com/matlabcentral/fileexchange/31926-fingerprint-minutiae-extraction. Accessed 3 Jan 2014.
48. Sagar VK, Alex KJB. Hybrid fuzzy logic and neural network model for fingerprint minutiae extraction. In International Joint Conference on Neural Networks, IJCNN. 1999; vol. 5, pp. 3255–9.

49. Schmid C, Mohr R, Bauckhage C. Evaluation of interest point detectors. Int J Comput Vis. 2000;37(2):151–72.
50. Harris C, Stephens M. A combined corner and edge detector. In: Alvey vision conference. Manchester, UK. 1988; vol. 15, pp. 50.
51. Rosten E, Porter R, Drummond T. Faster and better: a machine learning approach to corner detection. IEEE Trans Pattern Anal Mach Intell. 2010;32(1):105–19.
52. Mair E, Hager GD, Burschka D, Suppa M, Hirzinger G. Adaptive and generic corner detection based on the accelerated segment test. In Proceedings of the European Conference on Computer Vision (ECCV'10), September 2010.
53. Rublee E, Rabaud V, Konolige K, Bradski G. Orb: an efficient alternative to sift or surf. In IEEE International Conference on Computer Vision (ICCV). 2011; pp. 2564–71.
54. Li J, Allinson NM. A comprehensive review of current local features for computer vision. Neurocomputing. 2008;71(10):1771–87.
55. Lowe DG. Object recognition from local scale-invariant features. In: The proceedings of the seventh IEEE International Conference on Computer vision. 1999; vol. 2, pp. 1150–7.
56. Lowe DG. Distinctive image features from scale-invariant keypoints. Int J Comput Vis. 2004;60(2):91–110.
57. Lindeberg T. Scale-space theory in computer vision. Dordrecht: Springer; 1993.
58. Bay H, Tuytelaars T, Van Gool L. Surf: speeded up robust features. In Computer Vision–ECCV 2006, pp. 404–17. Springer; 2006.
59. Matas J, Chum O, Urban M, Pajdla T. Robust wide-baseline stereo from maximally stable extremal regions. Image Vis Comput. 2004;22(10):761–7.
60. Dalal N, Triggs B. Histograms of oriented gradients for human detection. In IEEE Computer Society Conference on Computer Vision and Pattern Recognition. CVPR 2005; vol. 1, pp. 886–93.
61. Jin L, Zhang TX. The generalization of moment invariants. Chin J Comput. 2004;5:011.
62. Deepika CL, Kandaswamy A, Vimal C, Sathish B. Invariant feature extraction from fingerprint biometric using pseudo Zernike moments. In Proceedings of the International Joint Journal Conference on Engineering and Technology. 2010; pp. 104–8.
63. Yang JC, Park DS. Fingerprint verification based on invariant moment features and nonlinear BPNN. Int J Control Autom Syst. 2008;6(6):800–8.
64. Chikkerur S, Cartwright AN, Govindaraju V. Fingerprint enhancement using STFT analysis. Pattern Recognit. 2007;40(1):198–211.
65. Hong L, Wan Y, Jain A. Fingerprint image enhancement: algorithm and performance evaluation. IEEE Trans Pattern Anal Mach Intell. 1998;20(8):777–89.
66. Maio D, Maltoni D, Cappelli R, Wayman JL, Jain AK. Fvc2002: second fingerprint verification competition. In Proceedings 16th International Conference on Pattern Recognition. 2002; vol. 3, pp. 811–4.
67. Jain A, Nandakumar K, Ross A. Score normalization in multimodal biometric systems. Pattern recognition. 2005;38(12):2270–85.
68. Shuai X, Zhang C, Hao P. Fingerprint indexing based on composite set of reduced sift features. In 19th International Conference on Pattern Recognition. ICPR 2008; pp. 1–4.
69. Gionis A, Indyk P, Motwani R, et al. Similarity search in high dimensions via hashing. In VLDB. 1999; vol. 99, pp. 518–29.
70. Rosdi BA, Shing CW, Suandi SA. Finger vein recognition using local line binary pattern. Sensors. 2011;11(12):11357–71.
71. Rosten E, Drummond T. Machine learning for high-speed corner detection. In Computer Vision–ECCV 2006. Springer, 2006; pp. 430–43.
72. Chikkerur S, Cartwright AN, Govindaraju V. K-plet and coupled bfs: a graph based fingerprint representation and matching algorithm. In: Zhang D, Jain AK, editors. Advances in Biometrics, volume 3832 of Lecture Notes in Computer Science. 2005; pp. 309–15. Springer Berlin Heidelberg.
73. Cappelli R, Ferrara M, Maltoni D. Minutia cylinder-code: a new representation and matching technique for fingerprint recognition. IEEE Trans Pattern Anal Mach Intell. 2010;32(12):2128–41.

74. Cappelli R, Ferrara M, Franco A, Maltoni D. Fingerprint verification competition 2006. Biom Technol Today. 2007;15(7):7–9.
75. Cappelli R, Ferrara M, Maltoni D, Tistarelli M. Mcc: a baseline algorithm for fingerprint verification in fvc-ongoing. 11th International Conference on Control Automation Robotics Vision (ICARCV). 2010; pp. 19–23.
76. Ferrara M, Maltoni D, Cappelli R. Noninvertible minutia cylinder-code representation. IEEE Trans Inf Forensic Secur. 2012;7(6):1727–37.
77. Cappelli R, Ferrara M, Maltoni D. Fingerprint indexing based on minutia cylinder-code. IEEE Trans Pattern Anal Mach Intell. 2011;33(5):1051–7.
78. Xu H, Veldhuis RNJ, Kevenaar TAM, Akkermans TAHM, Bazen AM. Spectral minutiae: a fixed-length representation of a minutiae set. IEEE Computer Society Conference on Computer Vision and Pattern Recognition Workshops, 2008. CVPRW '08. June 2008; pp. 1–6.
79. Xu H, Veldhuis RNJ, Bazen AM, Kevenaar TAM, Akkermans TAHM, Gokberk B. Fingerprint verification using spectral minutiae representations. IEEE Trans Inf Forensic Secur. 2009;4(3):397–409.

Lightning Source UK Ltd.
Milton Keynes UK
UKOW06n0329020515

250743UK00001B/12/P